Passive Components for Circuit Design

Passive Components for Circuit Design

Ian Sinclair

Newnes

OXFORD AUCKLAND BOSTON JOHANNESBURG MELBOURNE NEW DELHI

Newnes
An imprint of Butterworth-Heinemann
Linacre House, Jordan Hill, Oxford OX2 8DP
225 Wildwood Avenue, Woburn, MA 01801-2041
A division of Reed Educational and Professional Publishing Ltd

$\mathcal{R}$. A member of the Reed Elsevier plc group

First published 2001

© Ian Sinclair 2001

British Library Cataloguing in Publication Data
A catalogue record for this book is available from the British Library

ISBN 0 7506 4933 X

Typeset by David Gregson Associates, Beccles, Suffolk
Printed in Great Britain by Biddles, Guildford, Surrey

PLANT A TREE
British Trust for
Conservation Volunteers

FOR EVERY TITLE THAT WE PUBLISH, BUTTERWORTH-HEINEMANN
WILL PAY FOR BTCV TO PLANT AND CARE FOR A TREE.

Contents

Preface

Despite the very considerable increase in the use of ICs over the past ten years, passive components are still the mainstay of the electronics industry. The strong emphasis that is given to semiconductors, and ICs in particular, in teaching courses at all levels, however, causes the subject of passive components to be neglected, and many technicians would not know, for example, how to wind a 10 μH coil if they did not possess an amateur radio handbook to help them. Many, in fact, would not know how to specify or use such a component, now so much less commonly used. A passive component in this context has been taken to include any component which does not depend on the use of thermionic emission or semiconducting carrier effects, rather than the narrower definition of resistors and capacitors only. The simpler definition, that of no power amplification, would rule out components such as relays or cells that can usefully be classed as passive.

The comparative neglect of passive components leads to misuse of components. For example, resistors of an incorrect dissipation rating may be used in servicing, capacitors of high loss factor may be put into oscillating circuits and other similar problems of inappropriate selection arise which waste time and a very considerable amount of money. If anything, the intensive use of semiconductor devices, notably ICs, has increased the importance of the part that passive components have to play, because their use in feedback loops and bias chains means that the overall performance of a circuit depends on the passive components rather than on the active components.

This book is concerned with the main passive components; their fundamental action, parameters, variation with temperature, tolerances, stability, reliability and manufacturing methods and standards. The level of the book is aimed at a technician and engineering readership, both students and practising technicians and engineers, in industry or education. It is also useful background reading for less specialized courses in electronics, and for enthusiasts.

In the preparation of this book I have been greatly assisted by the provision of information from the electronics industry in general, and RS Components in particular, and I am most grateful to everyone who has contributed directly or indirectly in this way. I am particularly grateful to all at Heinemann Professional Publishing, particularly Matthew Deans, who by continual encouragement, have made this book possible.

I am most grateful also for the many helpful comments and suggestions made by Bob Pease of National Semiconductor Corp., Santa Clara, contributing editor of *Electronics Design News*, and author of *Troubleshooting Analog Circuits*, another title in the EDN series for design engineers (Butterworth-Heinemann).

Note that the term *ground* has been used throughout in preference to the older use of *earth* to mean connection to a common zero-voltage point. This has been done to make these references more intelligible to a wider readership. The convention of omitting full-stops in abbreviations such as ppm and rpm has also been followed.

Ian Sinclair
Spring 2000

Chapter 1

Fundamentals

For many decades, engineers have been able to understand and apply electrical principles on the basis of treating electricity as a form of fluid, and quantities such as *current* remind us of this view. Electronics, however, often forces us to consider in more detail what electric current *is* as well as what it *does*. All of the effects that we describe as electrical or electronic depend on units of electrical charge called electrons.

For our purposes we can regard electrons as being particles of unimaginably small size, and all repelling each other. This force of repulsion exists over distances that are large compared to the apparent size of the electron, and we attribute it to a quantity called charge (or electric charge). Each electron appears to carry exactly the same amount of charge, and by convention we take the sign of this charge to be negative for an electron. A charge is positive if it attracts electrons, and the size of any charge is measured by the force which it can exert on another charge at a measured distance. Electrons are the outer parts of any atom, and when an electron is removed from an atom the atom is left with positive charge, as an *ion*, attracting the electron back to it. These forces are, however, greatly changed when atoms are closely packed together as they are in most metals, and the result is to release electrons, allowing them to take part in carrying current.

Electric current

These materials which conduct electric current well do so because they contain large numbers of electrons which are free to move, and it is this movement of electrons (or any other charged particles) that we call electric current. Conversely, insulators are materials which, although they may contain vast numbers of electrons, retain these electrons strongly bound to the atoms so that none of them is free to move and so carry current.

- Conduction in crystals can also be due to mobile defects in the crystal, called holes, but these exist only within the crystal and cannot have any independent existence.

For any material in which movement of electrons is possible, the number of electrons that passes a fixed point per second is a measure of the strength of the current. To put numbers to this, a current of 1 A is registered when 6.25×10^{12} electrons pass per second – even this very small current demands that 6.25 million million electrons are moving past each second. In a good conductor, such as a metal, which contains vast numbers of free electrons, the average speed of the electrons is very low, of the order of a few centimetres per hour. In a poor conductor, such as a semiconductor material, the speed required for the same current would be considerably higher, measured in many centimetres per second.

- In a vacuum the electron speed is much higher, and it depends on the square root of the accelerating voltage. For example, in a cathode ray tube with accelerating voltage of 10 kV, the speed is around 60 million metres per second, but at 100 V acceleration the speed is around 6 million metres per second.

Since electric current is the movement of any particle that is charged, current flows when electrons move to and fro in a conductor, not only when the movement is in one direction only. Before anything was known of electrons, these two types of current were recognized and given the names AC (alternating current), and DC (direct current). In electronics, AC was at one time vastly more important, since all electronic signals consisted of AC, and DC was an incidental, used mainly for power supplies. Nowadays, many types of electronic signals consist of DC, so that we need to give equal weight to both of these types, although DC is often converted to AC so as to be more easily handled.

All electric current flows in a closed path, so that no atom is ever deprived of an electron because of the flow of electrons. An electron which leaves an atom will be replaced by an electron from another atom, so that electric current consists of a shuffling around of electrons, like musical chairs with no chairs being removed. The closer the chairs the less movement is needed, which is why the speed of electrons in a good conductor is low. Conversely, if the chairs are far apart, high speed is needed for the journey between chairs, corresponding to the high speeds of electrons in poor conductors.

Electrical quantities

The three electrical quantities that are of most interest to us are voltage, current and frequency. Current has been described already; it is the effect of the movement of electrons or other charged particles, and the strength of current is proportional to the amount of charge that passes a point per second. The voltage (more correctly *potential*, measured in units called volts) at a point is a comparative figure which measures the 'pressure' on electrons to move to or from that point to somewhere else. We usually take the potential (voltage) of the surface of the earth (a potential that is fairly constant) as that 'somewhere else', the zero level of potential, so that all voltages are measured by comparison to this zero, ground, or earth level. If the voltage of a point is such that electrons will flow to it from the earth, then that point has a positive voltage. Conversely, if electrons will flow from the point to earth, the point has a negative voltage. When the voltage at a point or the current in a circuit (a closed path) reverses at regular intervals then the voltage or the current is *alternating*, and the number of complete cycles of reversal (positive to negative to positive again, for example) per second is called the frequency.

The frequency of the AC power supply in the UK, and in most of Europe, is 50 Hz, where the hertz (abbreviated to Hz) is the unit of frequency that consists of one complete cycle per second. The standard frequency for North America is 60 Hz, and this has also been adopted for most of the American continent and in Japan.

- When a voltage is steady, it really ought to be referred to as ZF (zero frequency), because the term DC means direct current, and a phrase like DC voltage is really meaningless. A better choice would be constant current, but the use of DC voltage is now so common and so established that it has been followed in this book (and in most others) rather than the more appropriate ZF which could be applied to current or to voltage.

Active and passive components

Electrical/electronic circuits consist of complete closed paths for current, starting at one end with a source of voltage (EMF) which uses energy to pump electrons around the circuit, and ending back at the source. The analogy with a water pumping circuit is very close (Figure 1.1), and extends to the idea that turning off a tap at the tank (breaking the circuit) will make the water level rise, because of the pressure of the pump, in a vertical piece of pipe (the voltage rises when the circuit is disconnected). Inside the source of EMF (electromotive force, an old term) energy of some sort is

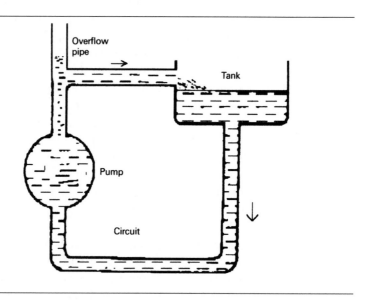

Figure 1.1   The similarities between a water circuit and an electrical circuit.

being used to push electrons around the circuit (when the circuit is connected) or to pile electrons up (when the circuit is disconnected).

Electronic circuits are built up by connecting components together with conducting paths which can be of metal wire, metal strips, or strips of other conducting materials such as doped silicon. In these paths, signals will be entering components and leaving components, and the power of a signal is measured by its voltage level multiplied by its current level. Electronic components generally are classed as being active or passive according to their effect on the power of signals applied to them. An active component can increase the power of a signal, using energy that is supplied in other ways, usually by a DC supply.

Passive components cannot increase the power of any signals applied to them and will almost inevitably cause power to be lost. Some passive components may increase the voltage of a signal, but this will be at the expense of current so that overall there is no gain of power. This definition is not totally watertight, because of the behaviour of varactor diodes and magnetic amplifiers, but is as near as we can get without becoming too elaborate at this stage.

A truly passive component can be used to reduce the power of a signal (deliberately), to select part of a signal by its voltage, its frequency or its time relationship to another signal, to change the shape of a waveform or to pass a signal from one section of a circuit to another; but in every case the power of the signal is decreased or unchanged, never increased. Resistors, capacitors and inductors are the fundamental passive components. There is a fourth type, however, called the gyrator, which is encountered in microwave circuits and which is of a more specialized nature.

An active component can increase the power of a signal and must be supplied both with the signal and a source of power. In many of the familiar active devices the source of power is a supply of current at a steady voltage, the DC supply, and the signal is fed in at one part of the active component and taken out from another. A few active devices have no separation of input and output, and some use an AC supply as the source of power and employ DC as their signals. The principle of using a power supply to increase the power of a signal, however, is unchanged.

An IC (integrated circuit) consists of a number of components, some purely active, some purely passive, others both active and passive, which are formed from the same materials, a semiconductor in single crystal or polysilicon 'chip' form. There may be terminals to this chip to which passive (or other active) components can be connected, but most of the circuit in the IC is inaccessible. Most ICs require a power supply, and are therefore classed as active components, and the advantage is that an IC can be treated as a single active component, with the reliability of a single component, whereas a circuit constructed from separate (discrete) components would contain huge quantities (ranging from thousands to millions) of components. This means that it would use a correspondingly large number of interconnections which would be considerably less reliable, since the reliability of a conventional circuit tends to decrease as the number of connections increases.

Thin-film circuits use techniques similar to those of IC manufacture in order to form passive components in miniature form, and thick-film circuits make use of materials other than semiconductors, such as metals that have been laid down by evaporation. Both thin-film and thick-film methods are used to make passive components for a variety of purposes.

Passive components

The fundamental main passive components are resistors, capacitors and inductors. The factor that they have in common is that they obstruct the flow of AC, although in very different ways, and are therefore said to present an impedance in the circuit. Resistors have a form of impedance which is termed resistance; and the effect of a resistance is to impede the flow of current equally for all frequencies of signal from DC upwards, although there are limits as we shall see later. In addition, when any current, AC or DC, flows through a resistor some of the energy of the current is converted to heat, causing the temperature of the resistor to increase. This heat is then passed on to the air (or anything else) that surrounds the resistor, and is a measure of the power that is being lost from the signal owing to the resistor. Wherever a resistor in a circuit passes current there will be a loss of power, and heat will have to be dissipated (usually to the air). The amount of heat will, however, be very small if the resistor is working at low voltage and current levels. At higher levels of power, if the heat is not dissipated, the temperature of the resistor will increase until the material melts or burns away. Usually this breaks the circuit, but for some resistor types it is possible for the material to fuse, decreasing the resistance, sometimes to short-circuit level.

Capacitors and inductors have *reactance* which, for a theoretically perfect capacitor or inductor, is not accompanied by resistance. Reactance also impedes the flow of current, but the amount of reactance is not a fixed quantity for one capacitor or inductor in the way that a resistor has a (more or less) fixed value of resistance. The reactance value of a capacitor or an inductor depends on the frequency of the current as well as on the component itself. At DC, capacitors have a very large reactance, amounting to complete insulation, and the size of the reactance decreases as the frequency of the signal is raised. Inductors have a very low reactance, almost zero for very low frequencies, but this value increases as the frequency of the signal is increased.

- The graph of reactance plotted against frequency for both capacitors and inductors will show minima and maxima due to self-resonance. An inductor will have self-capacitance and a capacitor will have some self-inductance, and in either case these will cause either series or parallel resonances.

When current passes through a pure reactance, high or low, no power is converted into heat and there is therefore no loss of power. In practice, no reactive component is perfect, although capacitors can come very close to the ideal, and any reactive component will have some resistance. When current passes, there will be a loss of power in this resistance, and the amount of such loss is expressed as a *power factor* (see later). The amount of loss should be very small for electronic components. Many circuits contain reactive components connected to resistors, and in such cases it is usual to assume that the power loss will be almost totally due to the resistor. Power loss in capacitors becomes significant for electrolytic capacitors subject to high ripple currents, and for other types mainly when signals of very high frequencies are applied.

The combination of resistance and reactance is known as *impedance* and the value of impedance for such a combination represents the total effect of resistance and reactance to an alternating current. We cannot calculate the value of impedance, however, by simply adding the value of resistance to the value of reactance. We have to

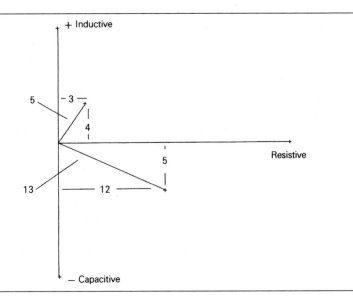

Figure 1.2 A phasor diagram with reactance plotted vertically and resistance horizontally. The combination of reactance and resistance has impedance whose value can be measured from the diagram as the distance from the origin (point of zero resistance and reactance) to the point that represents the combination of resistance and reactance. Two points are shown here with impedance values of 5 and 13 respectively.

find impedance in the same way as we can find distance between two points on a map when we have co-ordinates given. The similarity is illustrated in Figure 1.2, in which we map resistance, reactance (capacitive in this example) and impedance on a form of map called a phasor diagram. In this type of diagram, resistance values are always plotted along a horizontal scale and reactance on a vertical scale, down for capacitive reactance and up for inductive reactance.

Why do we plot the reactance along a line which is at right angles to the line of resistance? If we connect a resistance and a reactance in series and we pass an alternating current through them both, then by using an oscilloscope we can detect something that ordinary meters will not show us. If the oscilloscope can display two traces together, we can use one trace to show the voltage across the resistor and the other trace to show the voltage across the reactive component. These traces (Figure 1.3) are always out of step with each other, and the amount of this displacement is one

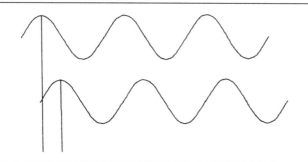

Figure 1.3 Sine waves 90° out of phase, as they would be seen on an oscilloscope.

quarter of a cycle of the wave, so that one waveform is reaching a peak as the other passes through its zero level.

Now for any repetitive action, like the voltage in a wave, we can represent a cycle by a complete circle (Figure 1.4). The angle that a radius of a circle sweeps out for a complete circle is 360°, so that one quarter of a circle is represented by a rotation of 90°. For this reason, the displacement of one wave relative to the other of one quarter of a cycle is called a 90° phase shift, and this is the reason for the title *phasor* diagram and the drawing of the value of reactance as a line at right angles to the line that represents resistance value.

What is 90° out of phase in such a circuit is the current through the reactive component compared to its voltage. The voltage across a resistor can be used as a measure of the current that flows through it, because a resistor does not shift the phase of a current. A reactive component does shift the phase of current as compared to voltage, however, and always by 90°, so that by drawing a line to represent reactance at 90° to the line for resistance, we can complete the 'mapping' to obtain a value for impedance which is of both the correct size (represented by length of line) and phase angle (angle to the horizontal). This phase angle will be less than that for a reactive component by itself, but when more than one type of reactive component is present the phase angle of the resulting impedance can change considerably as frequency is changed.

Capacitors differ fundamentally from inductors in the direction of the phase angle. For a capacitor, the wave of current is 90° before the wave of voltage; for an inductor the wave of current is 90° after the wave of voltage. The easiest way to remember this is in the word C-I-V-I-L (C, I before V, V before I in L) in which the

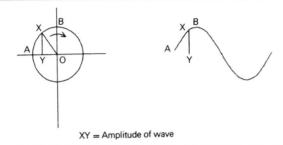

XY = Amplitude of wave

Figure 1.4 The connection between circular movement and a sine wave. The amplitude of the wave at any point corresponds to the distance from the horizontal axis of the circle to its rim for each point on the rim of the circle.

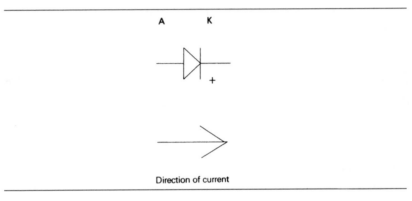

Direction of current

Figure 1.5 The diode and its symbol. The arrowhead of the diode symbol indicates the permitted direction of current through the diode.

standard symbol letters of C for capacitor, L for inductor are used as well as I for current and V for voltage.

In addition to these fundamental (and traditional) passive components, the semiconductor diode is also considered as passive. It can be classed as a form of resistor with low reactance, but whose resistance value to current in one direction is vastly greater than the resistance value for current in the opposite direction. The arrow in the diode symbol (Figure 1.5) is used to show the direction of current for low resistance.

• Some types of semiconductor diodes, notable varactor diodes and snap-turn-off diodes can behave as active components. Varactor diodes, for example, can be used as parametric amplifiers, but such use is outside the scope of this book.

Basic electrical facts and laws

The basic facts and laws that relate to passive components are:

1. **Ohm's law**, which states that for a metallic conductor at constant temperature, the resistance is constant. Since resistance is defined as the ratio of voltage across the component to current through it (Figure 1.6), Ohm's law means that for most resistors, a value of resistance measured at one current is valid for any other current value providing the temperature of the resistor does not change appreciably.

2. **Kirchhoff's laws**. The first law states that the current that leaves a circuit junction is equal to the current that enters it (conservation of current). The second law states that the sum of voltage across components in a circuit is equal to the circuit EMF (driving voltage). This is a law of conservation of voltage, and these two laws are simply another expression of the principle that electrons are never destroyed nor created. See Figure 1.7 for a diagrammatic explanation of these two laws.

3. The rules for addition of resistors or reactors in series and in parallel (Figure 1.8).

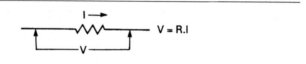

Figure 1.6 Definition of resistance. This is not a statement of Ohm's law, but our use of $V = IR$ depends on Ohm's law being obeyed.

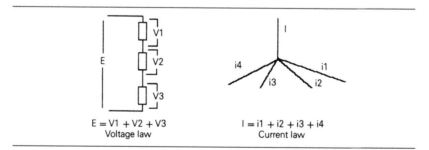

Figure 1.7 Kirchhoff's laws, which mean that there can never be any current or voltage unaccounted for in a circuit.

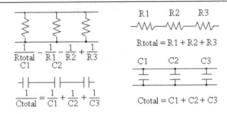

$$\frac{1}{\substack{\text{Rtotal} \\ \text{C1}}} = \frac{1}{\substack{\text{R1} \\ \text{C2}}} + \frac{1}{\text{R2}} + \frac{1}{\text{R3}}$$

$$\frac{1}{\text{Ctotal}} = \frac{1}{\text{C1}} + \frac{1}{\text{C2}} + \frac{1}{\text{C3}}$$

R1　R2　R3

Rtotal = R1 + R2 + R3

C1　C2　C3

Ctotal = C1 + C2 + C3

Figure 1.8　The rules for finding the effect of resistances, impedances and reactances in series and in parallel.

Reactance of inductor = 6.3 × f × L
where f = frequency in Hertz
　　　L = inductance in Henries.

f	10H	1.0H	0.1H	100mH	10mH	1mH	100μH	1μH
10Hz	630Ω	63Ω	6.3Ω	0.63Ω	Negligible values			
100Hz	6.3k	630Ω	63Ω	6.3Ω	0.63Ω			
1kHz	63k	6.3k	630Ω	63Ω	6.3Ω	0.63Ω		
10kHz	630k	63k	6.3k	630Ω	63Ω	6.3Ω	0.63Ω	
100kHz		630k	63k	6.3k	630Ω	63Ω	6.3Ω	0.63Ω
1MHz			630k	63k	6.3k	630Ω	63Ω	6.3Ω
10MHz	Impractical values				63k	6.3k	630Ω	63Ω

$$\text{Reactance of capacitor} = \frac{1}{6.3 \times f \times C}$$

f	1μF	1.0μF	100n	10n	1n	100p	10p
10Hz	1.6k	16k	160k	1.6M	16M	Too large	
100Hz	160Ω	1.6k	16k	160k	1.6M	160M	
1kHz	16Ω	160Ω	1.6k	16k	160k	1.6M	160M
10kHz	1.6Ω	16Ω	160Ω	1.6k	16k	160k	1.6M
100kHz		1.6Ω	16Ω	160Ω	1.6k	16k	160k
1MHz			1.6Ω	16Ω	160Ω	1.6k	16k
10MHz	Impractically small		1.6Ω	16Ω	160Ω	1.6k	

Figure 1.9　The formulae for reactance of capacitors and inductors, with tables of some values to show the effect of frequency. Note that reactance values for capacitors are negative (a consequence of the direction of phase shift).

4. The rules for finding the reactance value of capacitors and inductors, and how these quantities vary with frequency of signal (Figure 1.9).

5. Joule's law, which relates power dissipated by a resistive component with the current through the component and the voltage across it (Figure 1.10).

Joule's law:

Energy = V.I	Power = V.I.t
= I².R	= I².R.t
= V²/R	= V²t/R

Where V = voltage across resistor
I = current through resistor
t = time for which steady voltage and current
are sustained.

Figure 1.10 Joule's law which shows energy and power dissipated from a resistor. Note that a pure reactance (with no resistance) does not dissipate power.

Note that most of these laws and rules apply irrespective of the frequency of the signal, and the only effect of signal frequency is on the reactance value of reactive components.

Waves and pulses

When a signal voltage varies from one instant of time to the next in such a way that its various voltage levels repeat at definite and fixed intervals of time, then we are dealing with a wave or a pulse. The difference between the two is that the voltage level of a pulse is steady for most of its cycle, with only brief intervals in which the voltage changes rapidly. By contrast, the voltage level of a wave is continually changing and will be steady only for very short time intervals (short compared to the time for a complete cycle). For others, such as square waves, the level can be steady for much of the time of a wave, but then changes abruptly.

Some types of waves and pulses are generated as the natural results of the laws of physics. For example, the sine wave is the result of rotating a coil in the field of a magnet, which is the principle of the alternator. A pulse is generated from a different type of action, the switching of a voltage on and off at regular intervals. Other forms of waves and pulses are not generated by any natural means, and in general all of the waveforms that are used in electronic circuits are generated by electronic methods rather than by 'natural' methods. There are obvious exceptions, such as the use of devices such as switches and tacho-generators, but most electronic

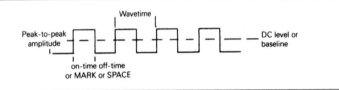

Figure 1.11 A square wave is produced by repeatedly switching a DC voltage on and off.

waves are generated by electronic circuits called oscillators or function generators.

- Because the sine wave is mathematically the simplest waveform and is (approximately) the one that is conveniently supplied to us by the domestic electricity board, it has been customary in the past to think of it as the most important waveform and to regard it as the source of all others.

Mathematically, this is true because any other waveform (wave or pulse) can be constructed by adding a number, sometimes a vast number, of sine waves which have related frequencies but different voltages and phases. Today, however, electronics makes very considerable use of square waves and pulses, and in many ways it is easier to follow the principles of waves with reference to a square wave than to a sine wave. It is also much easier to draw square waves!

The square wave, which is the wave produced by repetitive switching, is shown in Figure 1.11, together with some of the terms that are used to specify and describe waveforms and pulses. The first point to note in the diagram is that there is a voltage baseline around which the square wave is symmetrical. This is shown as a dashed line which in this case lies midway between the extreme voltages of the wave.

In many instances, this baseline is at ground voltage, but when it is not the wave is said to have a *DC component* whose size will be equal to the voltage between the baseline and ground voltage. This is the value which a moving-coil meter, or a digital voltmeter, set to a DC range would record if placed to measure voltage in the circuit, assuming that the presence of the meter did nothing to disturb the circuit conditions.

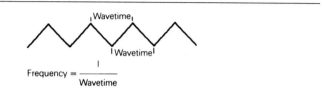

Figure 1.12   The period is the time between two corresponding points on successive waves.

The *amplitude* of a waveform is a figure expressed in units of volts. It may be the peak amplitude, which is the voltage difference between the baseline and one voltage extreme of the wave, or the peak-to-peak voltage which is measured from one voltage extreme to the other. These peak and peak-to-peak amplitudes are not directly measurable by a meter, and would be obtained from oscilloscope readings.

The *period* (or *wavetime*) is the time in seconds (more usually milli-, micro- or nanoseconds) between one point on the waveform and the next identical point on the next wave. The most convenient point for this measurement is the point where the voltage is at the baseline value, and we must also specify the direction of movement of voltage (positive to negative or negative to positive). It can, however, be just as easily measured between any other recognizable points, such as the places where the voltage starts to change suddenly (Figure 1.12). Once again, this is a measurement that is often taken by an oscilloscope, although a digital frequency/period meter is also useful. Of the two, the digital measurement is likely to be considerably more precise if precision is required.

The inverse of period (1/period) when the period is expressed in seconds is *frequency*, and the unit is the Hertz, the number of waves per second. Frequency is measured directly by digital frequency meters, and period can be calculated from it (period = 1/frequency) rather than the other way round. The quantity which is actually measured depends on the instrument which is available; period if an oscilloscope is used; frequency if a digital frequency meter is used.

For some waveforms, particularly pulses, the width may be important. A pulse is a form of square wave in which the on time is much less than the off time (Figure 1.13) and for such a wave it is useful to know for how long the voltage is at the higher level, the on time. This figure of pulse width is then used along with the

Figure 1.13 The width of a pulse is the time for which the important part of the pulse is at a steady voltage.

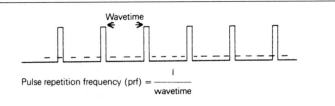

Figure 1.14 A train of pulses whose average value in this case is not zero – the average is marked by the dashed line.

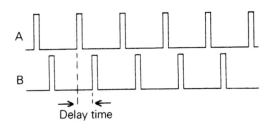

Figure 1.15 Illustrating a delay time between two square waves. This has to be measured between corresponding points.

figure of frequency or time between pulses. Pulses are referred to as narrow or wide, according to the width, and in general a narrow pulse will have a width of 1 μs or less. For digital equipment, pulse widths are usually measured in nanoseconds rather than in microsecond units. A train of pulses is an asymmetrical waveform, so that the average DC value is not midway between the peaks (Figure 1.14).

We can find that two square waves can have identical periods and amplitudes, but differ in time when they are displayed on a multiple-beam oscilloscope (Figure 1.15). The delay time is the difference in time between recognizable points such as the leading edges, the transition from negative to positive of each wave. This is the quantity that corresponds to phase shift of a sine wave, but the effects of the two have to be distinguished. The effect of delay on a square wave is simply to shift the time of the wave, preserving its shape. If there are phase shifts in a set of sine waves that make up a square wave, then unless the amount of phase shift is proportional to the frequency of each sine wave, the shape of the resulting wave will be changed.

Waveshape changes

The use of active components can carry out a large number of important changes to the shape of a waveform, but there are two particularly important changes which can be carried out by passive components, excluding diodes. These changes are called *differentiation* and *integration* and are particularly important when applied to pulse waveshapes. They are most easily described as they affect a waveform of square shape.

The square wave is made up of vertical (fast-changing) and horizontal (unchanging) portions. The effect of differentiation is to exaggerate the vertical portions and greatly reduce the amplitude of any horizontal portions. The overall amplitude may be greatly reduced during this process, so that differentiation often has to be followed by amplification. The effect of integration is to round off the vertical portions, smoothing out the waveform and reducing its amplitude so that, if carried to extremes, the output wave is almost DC (ZF), with just a slight trace of AC, called 'ripple'. Of all waveforms, only a sine wave does not change shape when it passes through either of these networks, although there will be changes in both amplitude and phase.

The quantity that is important for the purpose of either differentiation or integration is called time constant, often abbreviated to τ (Greek letter tau) and measured in milli- or microseconds. Where the differentiation or integrating circuit is made up from a capacitor and a resistor, as is often the case, the time constant is found by multiplying the values of resistance and capacitance. Using units of capacitance in nanofarads and resistance in kilohms, the time

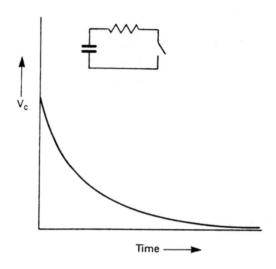

Figure 1.16 Capacitor charging. At the instant that the battery is connected, the capacitor starts to charge, but the charging is not instantaneous and is never totally complete.

constant will be in microseconds, and this set of units is usually more convenient than the textbook use of farads, ohms and seconds.

Suppose we imagine a capacitor connected in series with a resistor and a battery as shown in Figure 1.16. At the start, the capacitor is discharged and the battery is disconnected, so that the voltage across the capacitor is zero. What happens when the battery is connected? The answer is indicated in the graph – the capacitor will charge at a rate that starts fast but slows down as the charging proceeds. The shape of this graph is always the same, no matter what component values we use, but the amplitudes and time depend on the voltage of the supply and the values of resistance and capacitance that are used. The shape of the graph shows that the rate of charging decreases as the capacitor charges, so that, in theory, charging is never complete although for practical purposes the difference between the capacitor voltage and the supply voltage will be undetectable after some time. The simplest method of analysing what happens is by using the idea of time constant.

The time constant corresponds to the time that the capacitor takes to be charged to a voltage level that is about 63% of the final

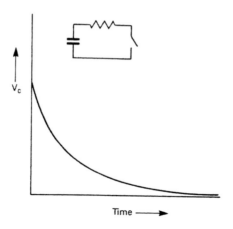

Figure 1.17 Capacitor discharging. At the instant that the switch is turned on, the capacitor starts to discharge, but this also is not instantaneous nor ever complete.

voltage, the supply voltage in this example. This may appear to be a very awkward figure to choose, but it is not a chosen amount, it is the consequence of the shape of the graph for the charging capacitor. The significant point is that after another interval of time equal to one time constant, the capacitor will have charged to about 63% of the remaining voltage, which means to a level equal to about 77% of the final voltage. After a time equal to four times the time constant, a capacitor will have charged to about 98% of the final amount, so that for all practical purposes, we can say that the capacitor is fully charged. For more exacting purposes, a time of 6τ or 7τ may have to be used to provide an accuracy of 100 ppm.

When a charged capacitor is connected across a resistor, the capacitor will discharge, and the graph of voltage plotted against time for discharging is shown in Figure 1.17. Discharging follows the same pattern as charging, with about 63% of the voltage discharged in a time equal to a time constant. As before, then, we can take it for many purposes that a capacitor will be completely discharged in a time equal to four time constants.

• Some types of capacitors, notably Mylar and high-K ceramic types, will need very much longer times for discharge because of voltage remanence or 'soakage' effects, see later.

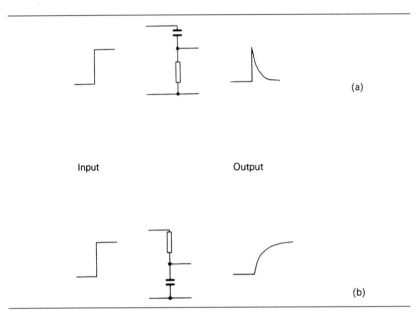

Figure 1.18   A voltage step and its effect on two CR networks. (a) The capacitor starts discharged, so that the voltage across it is zero and all of the voltage is across the resistor, decreasing as the capacitor charges. (b) The capacitor again starts discharged, but the output now is across the capacitor and will rise as the capacitor charges.

The significance of these charging and discharging time constants becomes apparent when we want to find out the effect of a capacitor-resistor circuit on a signal that consists of a sudden change of voltage. Such a signal is called a voltage step, and signals of this type are used in all types of digital circuits, particularly in computer circuits. A voltage step is the leading or trailing edge of a square wave.

   Figure 1.18 shows the shape of a voltage step from zero up to some fixed level, and the effect of the two possible simple capacitor-resistor circuits on this step. In each case, the waveform is the result of charging or discharging the capacitor, and the time that is needed is four time constants. In the first case (a), the capacitor is uncharged initially, so that each side is at the same voltage. After the step, however, the capacitor can charge, because one plate stays at the changed voltage and the other is free to pass current to ground through the resistor and so charge the capacitor. In the second case (b), the step voltage causes the capacitor to charge

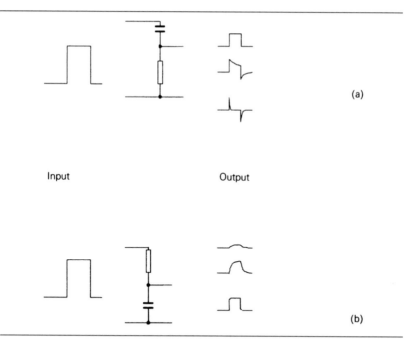

Input   Output

Figure 1.19 The effect of the CR networks on a square wave, for a variety of values of time constant. In each set, the top waveform is obtained for a large value of time constant relative to period, the lowest waveform from a small value of time constant relative to period.

through the resistor until the voltage across the capacitor equals the voltage of the step.

These changes in waveshape may not necessarily be noticeable if the step up in voltage is followed by a step down, forming a square pulse. As Figure 1.19(a) shows, if the time between the step up and the step down is very short, much less than the time constant of the capacitor and the resistor, then the shape of the square pulse is almost unchanged. When shorter time constants are used, however, the shape of the pulse is considerably changed, and with very short time constants, the square pulse is transformed into two pulses, one corresponding to each voltage step. This action of a time constant that is short in comparison to the time of a square pulse is differentiation, and the resistor-capacitor circuit is a differentiating circuit.

The integrator form of circuit, as illustrated in Figure 1.19(b), has very little effect on the square pulse if the time constant is very long compared to the time of the pulse. Using shorter time constants means that the capacitor is given less time to charge, so that the voltage never reaches the upper step level, and so the pulse is distorted into a slight rise and fall of voltage. Its effect is to smooth out a pulse, reducing the voltage step and extending the time. Both of these basic circuits are extremely important in all types of circuits that involve pulses, from TV circuitry to computer operation.

Differentiation and integration can also be unwanted effects. Sending a pulse through a long cable can have an integrating effect, smoothing out the pulse and spreading out the time for which the voltage changes. This effect was observed when the first transatlantic cables came into use, and was solved by matching the capacitance and inductance of the cable to resistors (terminating resistors) placed across each end. A properly matched cable will produce no integrating (or differentiating) effects. A full explanation of cable theory is beyond the scope of this book.

Undesired capacitance between two cables can result in a pulse on one cable producing a differentiated pulse on the other cable. In such examples, the capacitance is stray capacitance, and the resistance is the resistance that will exist in any circuit. One of the main problems in digital equipment is to maintain pulse shapes, particularly in a circuit that is physically large so that pulses have to travel along lines that are several centimetres in length. Many digital circuits make use of special circuits called drivers which are intended to reduce the problems of integration by supplying the pulse from a circuit that has very low resistance. This makes the time constant of this resistance with any stray capacitance very small, so as to have minimal effect on the pulse shape provided that the cable is correctly terminated. Once again, this topic is beyond the scope of this book.

Defects of square waves

We represent square waves on paper as being perfectly square, but a square wave, as examined by a good oscilloscope, may look anything but square. Some of the imperfections are due to the fact that the square wave must pass through networks (even if only the connections between the square-wave generator and the

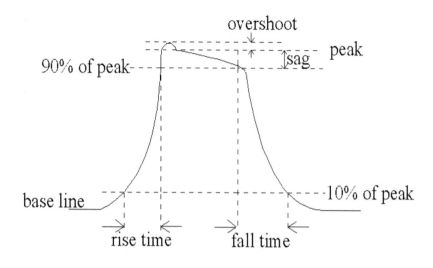

Figure 1.20   The imperfections that can exist in a square wave, caused mainly by stray capacitance and inductance.

oscilloscope) which will differentiate or integrate the wave. Other imperfections are due to the generator itself, caused by the imperfect switching action of active devices. Figure 1.20 shows in detail some of the imperfections that may exist in a square wave, and the same terms are used to describe these faults in any type of pulse.

The rise time measurement detects any integration of the wave. It is defined as the time taken for the voltage to rise from 10% of peak amplitude (measured from the base line, so that peak-to-peak is measured) to 90% of peak amplitude, expressed in microseconds or, more likely now, nanoseconds. The longer the rise time, the more the wave has been integrated, although square waves of very low frequencies can sometimes have quite long rise times (of the order of milliseconds) without detriment.

Overshoot is measured as the voltage of the overshoot peak as compared to the square wave high level, and is usually expressed as a percentage of that level. Overshoot is usually caused by inductance in the path of the wave, and is often followed by ringing, an oscillation of voltage which dies down after a few cycles. Both effects are very undesirable, and if inductance cannot be eliminated then it should be accompanied by enough resistance to ensure that the oscillation will be damped out even if the overshoot cannot be eliminated. In some cases, the overshoot will be due to inductance that has been added deliberately in order to improve the rise time of a circuit, and adjustment of the inductance value will be needed to ensure that the amount of overshoot and the rise time are both acceptable. When the overshoot is due to stray inductance, measuring the time of one cycle of overshoot can be a useful guide to estimating the amount of stray inductance.

The sag or droop of the wave is due to the presence of differentiating circuits in the wave path. A square wave of low frequency will have a flat top only if the circuit path is a DC path, containing no series capacitors and with no transformers used. At the higher frequencies, the differentiating effect of capacitors is less marked, but only a very small amount of differentiation is needed to produce a measurable amount of sag. Like overshoot, sag is quoted as a percentage of the normal steady high voltage level of the square wave.

The fall time, like the rise time, is measured as the time taken for the voltage to drop from the 90% level to the 10% level, and an extended fall time is also an indication that there is integration in the circuit. Many circuits use active devices in such a way as to make the rise time as short as possible, but with less attention paid to the fall time, so that it is normal to find that the fall time is greater than the rise time. A perfect square wave which has passed through totally passive circuits should have equal rise and fall times. Finally, the undershoot or backswing is measured and quoted in the same way as overshoot and may also be followed by ringing. This is less usual, even when inductance has been used to sharpen the rise time.

Metering and measuring

Practically all the applications for meters in electronics work are met either by some form of moving-coil meter or by a digital voltmeter. The moving-coil instrument consists of a magnet with pole-pieces shaped so as to generate a radial field (Figure 1.21) with a rectangular coil of wire suspended in the gap. The coil is supported by springs or by fine metal threads (in the taut-band version of the movement) which combine the actions of supporting the coil,

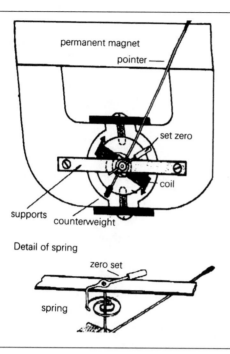

permanent magnet

pointer —

set zero

coil

supports

counterweight

Detail of spring

zero set

spring

Figure 1.21   The moving-coil meter movement, which relies on current passing through a coil to move the coil in a magnetic field.

passing current in and out, and restoring the coil to its rest position when it has been turned.

When current flows, the field of the magnet acts on the coil so as to turn it against the restoring torque (twisting effort) of the springs or threads. At some angle of turn the turning effort of the current equals the opposing effort of the springs, and the coil comes to rest so that a pointer attached to the coil will indicate a value on the scale of the meter. Typical sensitivities for modern instruments range from 1 μA to 50 μA for full-scale deflection (FSD), although movements in the 100 μA to 1 mA range are preferred for their low cost and rugged nature where the higher current is not a problem. In addition, the higher current movements have a lower resistance and a lower voltage drop across the movement at FSD.

Moving-coil instruments measure current, and any moving-coil movement will have its value of FSD current marked, along with the resistance of the coil. This allows the movement to be adapted to measure other current ranges and also to measure voltage and

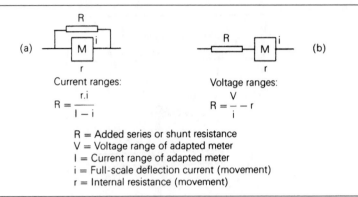

Figure 1.22 Extending the range of a current meter by adding shunt (a) and series (b) resistors to current and voltage ranges respectively.

resistance. The other current ranges are obtained by using shunts, resistors of low value which take a known fraction of the current from the meter so that the range indicated is greater than the natural range of the movement. Voltage is measured by adding resistors in series to the movement so that the desired voltage range will pass the correct amount of current through the movement.

Figure 1.22 is a brief reminder of the methods that are used to calculate values of simple series and shunt resistor values, although most meters use a combination of series and shunt resistance.

Resistance is measured by a moving-coil meter by setting the meter to read full scale with a known resistance value in series with a battery, and then connecting in the unknown resistance so as to cause the current reading to drop. The scale that is used is non-linear and is a reverse scale with the zero resistance mark at the position of maximum current and voltage, and the infinite-resistance mark at the position of zero current or voltage.

● Some older moving-coil instruments pass enough current, when switched to a low-resistance scale, to burn out delicate components.

Digital voltmeters operate on an entirely different principle. Referring to the much-simplified diagram of Figure 1.23, the voltmeter contains a precision oscillator that provides a master pulse frequency. The pulses from this oscillator are controlled by a gate circuit, and can be connected to a counter. At the same time, the

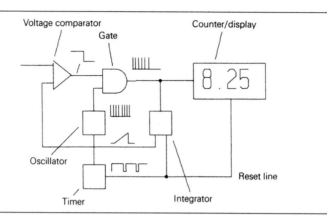

Figure 1.23 Principle of digital voltmeter. This depends on counting pulses that are added (integrated) until the resulting amplitude matches the input voltage.

pulses are passed to an integrator circuit that will provide a steadily rising voltage from the pulses. When this voltage matches the input voltage exactly, the gate circuit is closed, and the count on the display represents the voltage level. For example, if the clock frequency were 1 kHz, then 1000 pulses could be used to represent 1 V and the resolution of the meter would be 1 part in 1000, although it would take 1 s to read 1 V. The ICs that are obtainable for digital voltmeters employ much faster clock rates, and repeat the measuring action several times per second, so that changing voltages can be measured. Complete meter modules are available, containing an IC, display and some passive components. The simple system illustrated here is very susceptible to noise pulses in the measured input, and in a practicable instrument the input would also be integrated to reduce this effect.

- The example shows a type of circuit referred to as single slope, now obsolete. Modern digital voltmeters use a dual-slope action.

Like the moving-coil meter, the digital voltmeter can be adapted for other measurements. If the digital voltmeter is used, for example, to measure the (very small) voltage across a low-value resistor, then the meter can be scaled in terms of the current flowing, so that current measurements over a very wide range can be carried out. Resistance measurements are also possible, making use of a current regulator to pass a known current through the unknown resistor and measuring the voltage across the resistor. It is now possible to make at very low cost a digital meter whose precision is as good as the best moving-coil instruments of a decade ago, and the only penalty that is attached to the use of these instruments is that the reading can often take some time to settle down to a steady value, depending on the rate at which measurements are repeated.

Meter errors

There are two main sources of error in measurement made by any type of meter, the errors that are inherent in the meter itself, and those caused by adding the meter to a circuit. Of the two, the second type is often more important for moving-coil meters, because the precision of the meter is a known quantity (often 1% or better), whereas the effect on the circuit has to be calculated or estimated.

There are two basic methods for measuring current in a circuit. One is to break the circuit so that the meter, moving-coil or digital, can be connected in series, set at the highest current range. The circuit is then switched on and the current range of the meter is adjusted until a useful reading (in the range of $\frac{1}{4}$ to $\frac{3}{4}$ full scale usually) is obtained. The alternative is to measure the voltage across a small resistor ($1\,\Omega$ or less) which is permanently wired in series with the current circuit, so that a current measurement can be made without breaking any circuit. Either way, the resistance of the circuit is being changed by the addition of the meter.

The first method may be the only method available, but it is nearly always inconvenient to break the circuit, and it may be very undesirable to do so on a printed circuit board. In some circuits a clip-link may be provided so that current readings can be made easily. Such links can be placed at points in the circuit where the meter can be inserted without risk, because the insertion of a meter into points in the circuit where high-voltage pulses or large RF currents are present can cause damage to meters, and care must be taken to ensure that such waveforms do not pass through the meter. In general, a current meter should not be placed in any signal circuit. The arrangement of a permanent metering point in the form of a resistor in the current path is a better provision for current measurements, and if there is any possibility of unwanted signals the circuit can be arranged to ensure that these do not reach the meter by adding a bypass capacitor.

In general, inconvenient though the measurement of current may be, it disturbs a circuit much less than voltage measurement because the effect of current measurement is only to add a very small resistance in series with a circuit which will usually have a total resistance value that is very large in comparison. Voltage measurement using a moving-coil meter is much more likely to cause disturbance of circuit conditions. Although a voltage measurement is physically easy, involving clipping one meter lead to ground (or other reference point) and the other to the voltage point to be measured, the effect of the connection is to place the resistance of the voltmeter in parallel with the circuit resistance.

• This is no problem if a modern digital voltmeter with a very high input resistance (typically several kMΩ) is used, but older moving-coil instruments need to be used with considerable care.

Where the circuit resistances are high, comparable with the meter resistance, this can lead to readings which are quite unacceptably false. The meter reads as precisely as it can the voltage which exists, but this is not the voltage which would exist if the meter were not connected. The effect of a meter on a high-resistance circuit can be illustrated by considering a MOSFET circuit with a load resistor of 330 kΩ (Figure 1.24). Suppose that the meter is used on its 10 V scale, and that the MOSFET is cutoff so that the voltage ought to read 10 V. If the meter is made using a movement of 50 μA FSD, then for a 10 V reading on the 10 V scale, it would take 50 μA, so that the meter resistance is 200 k. The circuit, then, has now become a potential divider that uses a 330 k resistor (in the circuit) and a 200 k resistor (in the meter), and the voltage that now exists, and which is measured, will be $10 \times 200/530 = 3.77$ V.

This is an exaggerated case, with an unusually high load resistance value, but it illustrates just how far out a moving-coil meter reading can be. There can be no confidence in a voltage reading, therefore, unless the meter resistance is known and is much greater (at least ten times greater) than the resistance through which current will flow to the meter. The resistance of a digital meter is usually high; older types used a fixed value of around 10 MΩ on all

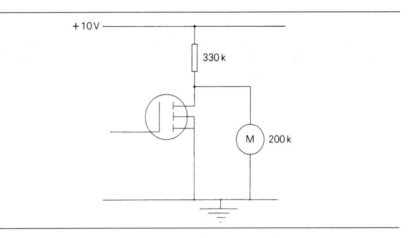

Figure 1.24   The effect of current through a moving-coil meter can cause low readings when the current has flowed through a large resistor.

ranges, but modern instruments provide very much higher values, several thousands of MΩ in some cases. The resistance of a moving-coil meter depends on the range that is used, and can be calculated from the figure of merit that is written on the dial. This figure is shown as kilohms per volt, and the meter resistance is equal to this figure multiplied by the full-scale range. For example, if the meter figure is 100 kΩ per volt, corresponding to a 10 μA full-scale movement, then on the 10 V range the resistance is $100 \times 10 = 1000$ kΩ, which is 1 MΩ.

The use of the oscilloscope causes the same form of disturbance as the voltmeter, with the difference that the oscilloscope is used mainly on points in the circuit where signal voltages are present. The signal voltage is usually at relatively low impedance, so that the introduction of the oscilloscope has less direct effect because of resistance. The stray capacitance of the oscilloscope, however, can alter the signal conditions, and the input resistance of the oscilloscope can cause alterations in the DC conditions which might in turn alter the amplitude or waveshape of the signal (by altering bias, for example). Most modern oscilloscopes are DC coupled and although older types have input resistances of 2 MΩ to 5 MΩ, modern units (mostly of Japanese manufacture) have standardized the input resistance at 1 MΩ with a capacitance of 25 pF to 30 pF. This allows the use of standardized resistive or FET probes, so that a 9 MΩ resistor can be used in a 10 MΩ input probe, and FET probes can provide input resistance levels of 1000 MΩ or more along with very low capacitance.

Functions of passive components

Modern electronic circuits in general contain drastically fewer numbers of passive components than their counterparts of ten or fifteen years ago, but the part that these passive components play in the action of the circuit has become correspondingly more important. These actions include:

1. Setting levels of current and voltage for DC and for signals.
2. Selection of waveforms (tuning and filtering).
3. Alteration of waveforms (integration and differentiation).
4. Making adjustments (volume controls, presets).
5. Transducing actions.
6. Environmental compensation (temperature mainly).

All six are important. The trend in recent years has been away from the use of simple passive components in favour of more complex devices. One example is the use of electroacoustic filtering circuits in place of inductor–capacitor filters. This, however, has not affected the importance of the fundamental passive components, particularly resistors, which are now made in an even greater variety than ever existed earlier. The emphasis now is on the reliability of such components which has now become even more important since each passive component is of increased importance. In the chapters that follow, we shall examine passive components in turn, considering their action, construction, and the reasons for selecting one type rather than another.

Chapter 2

Fixed resistors

A perfect resistor can be defined as a component which obstructs the flow of current equally at all frequencies and which releases heat in accordance with Joule's law:

$$W = V \times I$$

where W is the rate of dissipation of heat in watts, V is the voltage across the resistor in volts and I is the current through the resistor in amps.

There is no resistor that is perfect, and the main difference between a perfect resistor and any real resistor that can be manufactured is that no real resistor ever has a constant value for all possible frequencies, because signals at very high frequencies will flow around a resistor, through the stray capacitances, rather than through the material. For practical purposes, however, the definition is a reasonable one because we can think of stray capacitance as being something separate from the resistance of the resistor.

Resistors can be manufactured from a variety of materials, and some of these materials have been used for a very long time. In the early days of radio construction, amateur constructors would use a pencilled line on a wooden board as a high-value resistor, making connections by way of a bolt and washer in contact with an end of

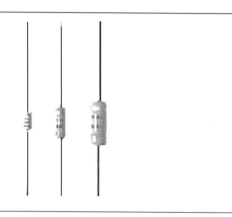

Figure 2.1 Modern resistors in $\frac{1}{8}$ W, $\frac{1}{4}$ W, and $\frac{1}{2}$ W sizes.

the line. This relied on the resistance of the mixture of graphite (a conductor) and clay (a non-conductor) that was used to make the pencil, and the harder the rating of the pencil (meaning a higher clay content) the greater the resistance of a line, given equal lengths and widths of lines being compared. Resistors of considerably lower value were made by winding insulated wire (using silk or cotton as insulators) around an insulated rod. Figure 2.1 shows a selection of modern resistors.

Carbon and metal are still the most important ingredients for the manufacture of resistors, but before we start to look at the practical aspects of resistor construction, specification and use, we need to know what it is about a material that controls its resistance. The most fundamental point concerns the materials that can be used, because it is the way that the atoms of these materials are constructed that gives rise to the ability to conduct electricity.

A material conducts electricity if the charged particles in the material can move. These particles can be electrons, holes, or ions. An electron is the smallest particle that we can trace inside the atom, and the outer layers of atoms consist of electrons. One picture of the atom that is helpful (although oversimplified) is of a small central portion, the nucleus, which is positively charged and surrounded by a number of electrons which are revolving around the nucleus, similar to planets moving around a sun. The number of electrons in a normal neutral atom is such that their combined negative charge balances exactly the positive charge of the nucleus.

In some materials (gases and liquids) atoms are fairly free to move and if electrons are added to atoms or removed from them, leaving the rest of the atom carrying an unbalanced charge, then the charged atoms will move when a voltage is applied. These temporarily charged atoms are called *ions*, and they are responsible for the conduction of electricity through gases, some molten materials and some solutions of materials in water. At some stage, an ion will regain or re-lose its odd electron and revert to its more normal and natural state, so that conduction by ions is not a permanent state, and it is not of major importance in electronics.

Conduction by movement of electrons and holes is much more important because it is the method of conduction in solid materials. The atoms in a solid material are much more closely packed together, and in the types of materials that we call crystals the packing follows a regular pattern. In the materials that we call metals, the packing of atoms in the crystals is so tight that some electrons, as many as one per atom, are not required to keep balance in the atom, because the packing has led to a sharing of electrons in which electrons are circulating so fast between atoms that any imbalance of charge has no time to have an effect.

In these materials, then, the stray *conduction* electrons are free to move in the material, making such materials good conductors. Another effect of the close packing of atoms in a crystal is that there will inevitably be imperfections. These can be caused by other types of atoms, either because no material can be perfectly pure or because the other atoms have been deliberately introduced in a process called *doping*. Alternatively they can be caused by faults, places where atoms have come together in a way that is not the closest possible packing. At an imperfection, it is possible to have gaps where electrons ought to be, holes in the otherwise perfect pattern of electrons. These holes, like the conduction electrons, can diffuse, meaning that they wander around the materials, and they act as if they were positively charged particles, rather more bulky than electrons and not moving so rapidly. Most metals conduct by a mixture of electrons and hole movement, although the best conductors, such as copper, conduct mainly by the movement of the conduction electrons.

There are, of course, many materials in which the atoms are not closely packed nor arranged into the patterns of crystals. These are non-conductors, insulators, and they have as important a part to

Table 2.1   Resistivity values for some common materials

Values are in units of nano-ohm. metres. To find
absolute values, multiply by 10^{-9}. Values for pure
metals are shown first, followed by values for alloys.

Metal	Resistivity
Aluminium	27.7
Copper	17.0
Gold	23.0
Iron	105
Nickel	78.0
Platinum	106
Silver	16.0
Tin	115
Tungsten	55.0
Zinc	62.0
Carbon-steel	180
Brass	60
Constantan	450
Invar	100
Manganin	430
Nichrome	1105
Nickel-silver	272
Monel metal	473
Kovar	483
Phosphor-bronze	93
18/8 stainless steel	897

play as the conductors. Most of the non-metals are insulators, although several are semiconductors in which a small alteration of the structure (caused by heating, light or the addition of other materials) can alter the conductivity to an enormous extent.

The contribution that the material itself makes to the resistance of a sample is measured by its *resistivity*. The dimensions of the material have their own contribution to make, and the resistance of a sample of material will depend on its length (the longer the sample the greater the resistance) and inversely on its area of cross-section (the greater the area of cross-section the lower the resistance), The resistance of any piece of material is therefore found from the formula:

$$R = \frac{\rho l}{a}$$

in which the quantity ρ (Greek rho) is resistivity, l is length, and a is the area of cross-section (assumed uniform).

Resistivity is a constant for a material, not affected by dimensions, and a few typical values of resistivity are shown in Table 2.1. The units of resistivity are ohms × metres (ohms multiplied by metres, not ohms per metre) and metals have values that are very small, of the order of 10^{-8} Ωm. Insulators have values that are very high, of the order of 10^{10} Ωm or more, making resistivity the most striking and obvious measure of the differences between these types of materials.

- A few materials, notably some metals at temperatures close to absolute zero (0K), have zero resistivity and hence zero resistance. These are superconductors, and are beyond the scope of this book.

Resistor characteristics

The two most obvious characteristics of any fixed resistor are its nominal resistance value and its maximum dissipation. The nominal resistance value will be colour coded (Table 2.2) or printed on to the body of the resistor, and in the UK the value will be shown using the convention that is explained in Figure 2.3.

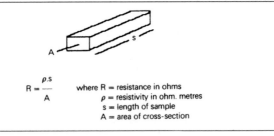

$$R = \frac{\rho.s}{A}$$
where R = resistance in ohms
ρ = resistivity in ohm. metres
s = length of sample
A = area of cross-section

Figure 2.2   Finding the resistance of a sample of material whose resistivity value is known.

Table 2.2   The values of the colour codes

Colour	Figure	Colour	Figure
black	0	green	5
brown	1	blue	6
red	2	violet	7
orange	3	grey	8
yellow	4	white	9
silver	0.01 (3rd band only)	gold	0.1 (3rd band only)

For tolerance coding, silver $= 10\%$; gold $= 5\%$

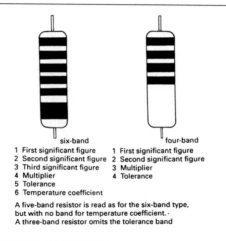

six-band   four-band
1 First significant figure   1 First significant figure
2 Second significant figure 2 Second significant figure
3 Third significant figure   3 Multiplier
4 Multiplier   4 Tolerance
5 Tolerance
6 Temperature coefficient

A five-band resistor is read as for the six-band type,
but with no band for temperature coefficient. ·
A three-band resistor omits the tolerance band

Figure 2.3 The colour coding methods that are using three, four, five or six colour bonds.

Other forms of coding are also used, particularly for metal-film resistors of US manufacture.

The maximum dissipation is not usually marked on to the resistor in any way, but the specification of the resistor will include its maximum dissipation, using values such as 0.125 W, 0.25 W, 0.5 W and so on. The fundamental unit of resistance is the ohm, defined as the ratio of voltage across the resistor to current through it. The fundamental unit of dissipation is the watt, with power in watts defined as the product of voltage and current, $V \times I$.

> The letter R is used for ohms; K and M have the meanings of kilohm and megohm respectively. The letter is placed where the decimal point would normally occur. A tolerance letter follows using $F = 1\%$, $G = 2\%$; $J = 5\%$; $k = 10\%$; $M = 20\%$

Table 2.3 The BS 1852 convention used for printing resistance values. This dispenses with the decimal point and hence avoids mistakes due to misreading a dot

R33 = 0.33 Ω	2R2 = 2.2 Ω	470R = 470 Ω
1K2 = 1.2 kΩ	22K = 22 kΩ	4M7 = 4.7 MΩ
5K6G = 5.6 kΩ, 2%	33Kk = 33 kΩ, 10%	

Ohm's law, noted briefly earlier, is one of the most fundamental and certainly the most misquoted of all electrical laws. The formula $V = R \times I$ is *not* an expression of Ohm's law, simply a statement of the definition of resistance. Ohm's law states:

The resistivity of a metallic conductor is constant at a constant temperature.

The implication is that if we obtain a value of resistance from one measurement of voltage and current, then this ought to be the same as the value that we would obtain using any other voltage or current value, assuming that the temperature has not changed. In other words, the resistance at a constant temperature is determined by the material and its dimensions, not by the amount of current or voltage.

This is certainly not true of semiconductor junctions, and not necessarily true of intrinsic semiconductors, although the resistors made by diffusion in ICs are as linear as many metal-oxide types. A semiconductor which has a voltage V across it when a current I flows will have a resistance value of V/I which will not be a constant if the temperature of the semiconductor is altered even slightly, since semiconductor resistance values are affected much more by temperature than metals. Ohm's law makes it possible to carry out voltage and current calculations without reference to charts or to elaborate formulae. If Ohm's law did not apply to any material, it would be impossible to mark a resistor with a value of resistance. At best, we would have to mark it with the value of current (or voltage) at which this value was measured, but we would not know what resistance value to use at any other value of current or voltage.

The resistance value that is marked by colour code or in numbers on the body of a resistor is not, however, the exact value that you might measure for that resistor. For most of our applications of resistors we need to know only the approximate value of resistance, so that the actual value will differ from the marked value by an amount that we call the tolerance, and which can be positive or negative. The tolerance could be marked in ohms, and for a few specialized wire-wound resistors you can find values such as $5R \pm 0.1R$ marked, but it is much more usual to find the tolerance expressed simply as a percentage.

A tolerance of 10%, for example, means that the actual value of the resistance can be expected to be anywhere between 10% over and 10% under the marked value. At one time, 20% was a normal tolerance level, but modern methods of manufacturing resistors have brought this down to 5% or 2%, and much closer tolerance values such as 1% or 0.5% can be found for specialized purposes (such as in measuring instruments). Tolerance values express the limitations of the manufacturing process, and are not connected with the changes of resistance that can take place during storage, during use, and because of changes in temperature through self-heating or changes in the surrounding (ambient) temperature.

Preferred values

When resistors are manufactured, certain preferred values are aimed at. These preferred values are based on a scale that starts at one and which has each step in a geometrical progression, meaning that each value is some constant factor times the previous value. In practical terms, this means that each number in the scale will be the same percentage higher than the previous one, allowing for rounding the values to a whole number. In such a series, the values can be arranged so that all possible values are within the tolerance range of one or two of the preferred values in the series.

• Modern manufacturing methods make it easier now to achieve a very close tolerance, and most of the existing tolerance series reflect the older manufacturing.

For example, if we take the old 20% tolerance series, each number in the series is a sixth root of ten above the previous one. The sixth root of ten has a value which is approximately 1.46, and if we round off numbers in a series that starts at 1, then we get the familiar set of numbers 1, 1.5, 2.2, 3.3 and so on, with each number approximately 40% higher than the previous one in the series, if we have a value which is between two preferred values, then whatever this value happens to be, it must fit at least one preferred value. For example, if we have a resistor of value 1K8 which is between the values of 1K5 and 2K2, then this value is 20% above 1K5 and 18% below 2K2, so that such a resistor could be marked as either a 1K5 or a 2K2 and still be within the limits of 20% tolerance. The 10% series uses as its multiplier the twelfth root of ten,

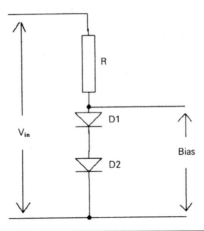

Figure 2.4 A circuit in which the tolerance of resistance makes very little difference to the bias voltage across the diodes, providing the temperature is maintained constant.

with a value of approximately 1.21, and the step is about 20%; the
5% series uses the twenty-fourth root of ten to give a factor of
about 1.1, so that the step is 10% and so on.

The overall effect is that if a manufacturer aims to make resistors
of the preferred values in a series, then any resistor that is manufactured must fit somewhere in the series. More to the point, a given
batch of resistors will nowadays often have quite a small spread of
tolerances, so that a set of 10% resistors are unlikely to contain
many whose tolerance is closer than 10%. At one time, manufacturers simply accepted large spreads in tolerance levels, and sorted
out resistors into batches according to the tolerance level of their
values; but this practice has now died out. Table 2.4 shows the
20%, 10% and 5% series of numbers which are used for all values
– the 5% range is used also for the 1% resistors, and a six-band
colour coding is also used for very close-tolerance resistors. Higher
or lower values are simply multiples of these numbers, so that we
can have 1R5, 15R, 150R, 1k5, 15k, 150k, 1M5 and so on from the
number 1.5 in the series.

● Modern manufacturing techniques can now ensure that the majority of resistors in a batch are of almost exactly the marked
  value, so that large tolerances are a thing of the past.

Table 2.4 The 20%, 10%, 5% and 1% tolerance series of preferred values. The
20% tolerance ranges are seldom used nowadays

Set	Values
E6 20%	1.0 1.5 2.2 3.3 4.7 6.8
E12 10%	1.0 1.2 1.5 1.8 2.2 2.7 3.3 3.9 4.7 5.6 6.8 8.2
E24 5%	1.0 1.1 1.2 1.3 1.5 1.6 1.8 2.0 2.2 2.4 2.7 3.0 3.3 3.6 3.9 4.3 4.7 5.1 5.6 6.2 6.8 7.5 8.2 9.1
E96 1%	1.00 1.02 1.05 1.07 1.10 1.13 1.15 1.18 1.21 1.24 1.27 1.30 1.33 1.37 1.40 1.43 1.47 1.50 1.54 1.58 1.62 1.65 1.69 1.74 1.78 1.82 1.87 1.91 1.96 2.00 2.05 2.10 2.15 2.21 2.26 2.32 2.37 2.43 2.49 2.55 2.61 2.67 2.74 2.80 2.87 2.94 3.01 3.09 3.16 3.24 3.32 3.40 3.48 3.57 3.65 3.74 3.83 3.92 4.02 4.12 4.22 4.32 4.42 4.53 4.64 4.75 4.87 4.99 5.11 5.23 5.36 5.49 5.62 5.76 5.90 6.04 6.19 6.34 6.49 6.65 6.81 6.98 7.15 7.32 7.50 7.68 7.87 8.06 8.25 8.45 8.66 8.87 9.09 9.31 9.53 9.76

Changes in value

Changes in resistance value affect both the designer of circuitry and
the service engineer. The designer must decide what effects the
maximum change in value will have on the circuit and whether
this will be significant. The service engineer will have to determine
whether a measured value for a resistor is within tolerance and if it
is not, whether the change can account for a fault in the circuit. It
is important to realize that changes in resistance value may not
necessarily be serious, even if the change is quite large.

For example, Figure 2.4 shows a circuit in which a bias voltage is
obtained by passing current through a pair of diodes. The amount
of current is determined by the value of a resistor R, whose marked
value is 3K3. Now if the value of the resistor is measured and
found to be 2K2 or 4K7, is this serious? The change is certainly
beyond the limits of tolerance, but the effect on the bias voltage is
very small, because semiconductor diodes do not obey Ohm's law,
and a change of current does not cause a change of voltage that is
in proportion. A change of temperature, however, will alter the conditions considerably, but that's another story.

On the other hand, there are circuits in which a change in the
value of a resistor would cause a very large change in the circuit
conditions – to such an extent that such circuits are not used if at

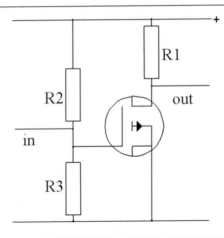

Figure 2.5 An FET circuit which is never used because of its dependence on exact resistor values.

all possible. Figure 2.5 shows a FET gate biased by a resistor chain, but with the source grounded. In this circuit, a small change in the value of either biasing resistor would cause a change of gate voltage that would seriously affect the biasing of the stage, and for that reason such a biasing circuit is never used. A corresponding bipolar transistor circuit is also troublesome because of the variation of the h_{fe} between transistors, an effect that is much more serious than variations in resistor values.

Changes in resistance value occur for a variety of reasons, of which the most significant is changes of temperature during operation, dealt with later in this chapter. There are also changes brought about as a result of soldering, as a consequence of damp and hot storage conditions, and also as an effect of age. The precise levels of change depend on the method of manufacture (see later), but for the old types of carbon composition resistors the change on soldering can be as large as 2%, the effect of damp heat can be to cause a change as large as 15%, one year of shelf life can cause a change of up to 5%, and one year of working conditions at 70°C and full rating can result in a change of up to 15%. Since the resistor might have started at 20% over or under its marked value, you can see that the value which could be measured after some time in service could have changed quite considerably without implying a fault in the resistor. More modern types of resistors exhibit much smaller changes in resistance value, but the guaranteed maximum drift figures are often higher than the amounts found on test.

Maximum dissipation

The maximum dissipation of a resistor is the amount of electrical power which it can continually change to heat without damage to the resistor. When a resistor dissipates power, it becomes hotter than its surroundings so that heat can flow from the resistor to its surroundings. The lower the temperature of the surroundings, the greater the amount of heat a resistor can dissipate because heat will flow more easily when the difference in temperature between the hot object and its cold surroundings is large. It would be very unusual for a resistor to be kept permanently in cool surroundings, however, and the dissipation of resistors is usually quoted for surroundings at 70°C. This does not imply that resistors cannot be used above this temperature, and most resistors can be used to temperatures of 100°C or more, but the dissipation rating of a resistor used at these higher ambient temperatures will have to be reduced.

Manufacturers tend to be cautious about revealing the extent of derating that might be needed, but a reasonable rule of thumb is to halve the dissipation rating for each 30°C rise in ambient temperature above the 70°C mark. This applies only to a limited temperature range – you cannot, for example, assume that you can run a resistor at high temperatures even at a very reduced dissipation rating.

For military equipment, the derating of resistor dissipation values will be specified, and this is often true also for other assemblies which have to conform to tight reliability specifications (civil aircraft, for example). Commercial uses of resistors in domestic equipment such as TV receivers often allow for little or no derating, but even for these uses it is becoming more common now to make some allowance.

Excessive dissipation over a long period can result in the resistor itself, the printed circuit board (PCB) or neighbouring components appearing blackened or blistered. If the neighbouring components are capacitors, this may lead to a fault occurring which would not necessarily be attributed to a resistor running hot, so that it is always good practice to examine all components carefully in the region where a fault has been traced. Resistors which are known to dissipate a large amount of power will often be mounted on metal pillars or on long leads, making a longer path for heat to travel to the PCB and also raising the hot resistor above the general level of other components and so allowing it to dissipate heat more freely to the air around it. Another technique is to clamp the resistor to a large area of copper foil that acts as a fin. It is possible, however, for a resistor to be damaged by a comparatively short duration overload which makes the internal temperature of the resistor high enough to cause damage without lasting long enough to make a visible change to the exterior.

The quoted temperature for a resistor rating of 70°C may seem quite high (water boils at 100°C), but this can easily be reached in a crowded circuit layout inside a casing, and particularly if the circuit contains several high dissipation resistors, power supplies, or thermionic components (such as CRTs, or magnetrons). Ideally, each PCB in a circuit should have a free flow of air around each surface, and this should be boosted by the use of a fan if such a free flow cannot be guaranteed. Although the designer of equipment can specify the positioning of PCBs and the design of cabinets, the end-user is under no constraints, so that you can find electronic equipment such as computers or TV receivers placed against radiators or close to hot-air vents. Even the most pessimistic assumptions that the designer makes in respect of ambient temperatures may turn out to be optimistic in view of the way that the user sites and treats the equipment.

It is often useful to know approximately what temperature rise will occur in the body of a resistor when it is used at its maximum (or any other) rating. There is no simple way of calculating such temperature rises, because they depend so much on the way that the resistor has been constructed, and on factors like the surface finish. Figure 2.6 shows charts, based on fairly old measurements on carbon resistors, which show the temperature rises that can be expected with dissipation for resistors rated at 0.5 W and 1 W respectively. These indicate a rise of around 60°C above ambient temperature for full rated dissipation, so that if the ambient temperature were the usual 70°C figure that is quoted for rating purposes, the resistor surface temperature could be as high as

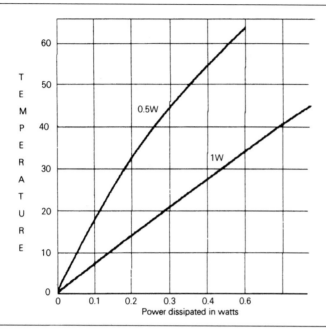

Figure 2.6 Graphs which give an indication of the temperature rise for a resistor with different dissipation values.

130°C. This is only an approximate figure, because the ability of a surface to dissipate heat increases as its temperature is raised, but it can be a useful guide which is often very difficult to obtain.

Temperature coefficients

All resistor types exhibit a change of resistance value when temperature changes, and in this section we are concentrating on temporary changes only, assuming that the resistance value will return to normal at the normal temperature. Temperature has two effects on resistance. One effect is caused by alterations in the size of the resistive material, whose length and area of cross-section will both increase. The other effect is the change in the value of resistivity, and for all of the materials that we use for manufacturing resistors, the resistivity change is so much greater than the changes due to

The temperature coefficient α is defined
by the equation:
$$R_\theta = R_0(1 + \alpha\theta)$$
in which R_θ is resistance value at temperature θ
and R_0 is the resistance at 0°C.
If resistance R1 is known as temperature θ1 then
to find value R2 at temperature θ2, use:

$$R2 = R1\left(\frac{1 + \alpha\theta2}{1 + \alpha\theta1}\right)$$

Figure 2.7 How temperature coefficient is defined. The sizes of temperature coefficient are often quoted in parts per million (ppm) rather than as a fraction.

the alteration in the dimensions that we can concentrate on this change alone.

The quantity that is used to measure the effect of temperature changes is called the temperature coefficient, and strictly speaking the coefficient that we use is the temperature coefficient of resistance. The definition, illustrated in Figure 2.7, is the fractional change in resistance per degree Celsius. If the resistance increases as the temperature increases, then the temperature coefficient is positive (a PTC resistor), and if the resistance decreases as the temperature increases then the temperature coefficient is negative (an NTC resistor). In general, good conductors such as metals always have a positive temperature coefficient, but semiconductor materials normally have a negative coefficient, and carbon resistors can have a negative coefficient also. Table 2.5 shows the values of temperature coefficient of resistivity for some materials; because the effect of changes of dimension is very small, these figures are for all practical purposes identical to the temperature coefficient of resistance figures.

The values of temperature coefficients as quoted in many data books are very small, using the definition of the coefficient as fractional change per degree Celsius. It is more common to make use of temperature coefficients quoted as parts per million per degree in order to provide more accessible and memorable figures. For example, rather than quote the temperature coefficient of a resistor as being $2.5 \times 10^{-4}/°C$ we write it as 250 ppm/°C, a more compact form and one that is much easier to remember and work with. Table 2.6 illustrates how the different forms used to quote these values can be converted from one to the other. For resistors

Table 2.5 A table of temperature coefficient of resistivity for some selected materials; this is virtually the same value of temperature coefficient of resistance

Values are in units of 10^{-3} K^{-1}. To find
absolute values, multiply by 10^3. Values for pure
metals are shown first, followed by values for alloys

Metal	T/Coefft.
Aluminium	4.2
Copper	4.3
Gold	3.6
Iron	6.5
Nickel	6.5
Platinum	3.9
Silver	3.9
Tin	4.3
Tungsten	4.9
Zinc	4.1
Carbon-steel	6.3
Brass	2.0
Constantan	0.2
Manganin	0.02
Nichrome	0.017
Nickel-silver	0.026
Monel metal	2.0
Phosphor-bronze	3.0

Table 2.6   Converting between ppm/°C values and the fraction/°C values

Conversions

From	To	Multiply by
K^{-1}	ppm/°C	10^6
Table	ppm/°C	10^3
ppm/°C	K^{-1}	10^{-6}

Note: 'Table' refers to figures in Fig. 2.5
10^6 means one million, 1 000 000
10^3 means one thousand, 1 000
10^{-6} means one millionth, $\frac{1}{1\,000\,000}$
K^{-1} is the standard unit of temperature coefficient

constructed from metal these coefficients will all be positive, but some carbon resistors can exhibit negative coefficients.

Noise in fixed resistors

In all substances, even at the very low temperatures of liquid helium (−270°C), the electrons and other parts of the atom are vibrating and the extent of the vibration increases as the temperature is increased. The movement of electrons constitutes an electrical signal and since the vibration is random and not at any fixed frequency the signal that is produced is electrical noise. The predominant noise effect in resistors is called *Johnson* or *thermal* noise.

Noise in active devices, where electrons are arriving randomly at electrodes, is usually more important than noise in passive devices, but the noise in some resistors, notably the obsolescent carbon-composition types with DC flowing, is quite high because of the material and the construction of the resistor. Any amplifier of high gain must have a low-noise first stage, since any noise at the input will be subject to the full gain of the amplifier, and any resistors that are used in this first stage will be specified as low-noise types. The most critical resistance in any amplifier circuit is the input resistance, and the lower the value that can be used the better. The converse is true for a load resistor, because the gain obtained from a high value of resistance outweighs the higher noise figure.

● The worst offenders from the point of view of noise are likely to be carbon composition resistors with DC flowing through them.

Resistors that are wire-wound or which make use of metal films have lower noise values than other types, and should be specified for the input stage of low-noise amplifiers. The actual amount of noise voltage rises as temperature is increased, so that cooling of input stages is extremely useful in obtaining low-noise performance. The signal bandwidth is also important, because the smaller the bandwidth that a signal uses the lower is the noise in any resistor through which the signal flows. Noise level for a resistor is usually quoted as microvolts of noise per volt of applied voltage for a standard 1 MHz bandwidth. Figure 2.8 shows the equation for resistor thermal noise.

The value of Boltzmann's constant is 1.38×10^{-23}. Do not confuse this with the Stefan–Boltzmann constant used for thermal radiation calculations.

In more practical units, we can calculate the noise voltage in nanovolts as:

$$7.43\sqrt{RT\Delta f}$$

where the units are kilohms, Kelvin and kHz respectively. We can take the Kelvin temperature as $°C + 273$.

$$E = \sqrt{4.R.k.T.\Delta F}$$

where E is RMS noise signal voltage in volts
R is resistance value in ohms
k is Boltzmann's constant
T is temperature in Kelvin
ΔF is bandwidth in Hz

A more practical form gives the noise
voltage (RMS) in nanovolts as:

$7.43 \sqrt{R.T.\Delta F}$ when the units
are kilohms and kHz for resistance
and bandwidth respectively.
To find temperature in K, add 273 to
the Celsius temperature.
Example: Find the RMS noise in a resistor of 47K carrying
a signal of 1 MHz bandwidth at 30°C.
Using the simplified formula, this gives:

$E = 7.43 \times \sqrt{47 \times 303 \times 1000} = 28038\,\text{nV}$
which is 28.04 μV (rounded)

Figure 2.8 How noise in a resistor is related to resistance value, as well as to temperature.

Example: Find the RMS noise in a 47 kΩ resistor carrying a signal of 1 MHz bandwidth at 30°C. Using the simplified formula:

$$E = 7.43 \times \sqrt{47 \times 303 \times 1000} = 28\,038\,\text{nV or } 28\,\text{mV (rounded value)}$$

High frequency effects

A resistive material should have a constant value of resistivity, unaffected by the frequency of the signal currents passing through it. This is basically true of most materials, but all practical resistors will have stray capacitance across their leads, and because of this stray capacitance the resistor will have an impedance value (resistor in parallel with capacitor) which will decrease as the signal frequency is increased. The amount of stray capacitance depends on the shape and size of the resistor and there is no simple guide to the maximum frequency at which a resistor can be used. Over a large range of the higher frequencies, the stray inductance of the resistor will also play a part, so that the resistor may appear to have a resonance frequency at which its apparent resistance is at the rated value. For some UHF circuits, stray inductance can be more important than the stray capacitance when the applied frequency is higher than the self-resonance frequency, and leadless resistors (SMC) have to be used. The Boella effect (see later) rules out the use of carbon composition resistors for high-frequency applications.

Voltage effects

The maximum working voltage for a resistor should never be exceeded, even for a short period. This voltage can be as low as 150 V for a 0.25 W resistor, rising to as high as 1000 V for a thick-film type. One effect of overvoltage is to cause overheating if sufficient current can flow, but apart from this, overvoltage can cause problems in its own right. Excessive voltage can give rise to sparking over parts of the resistive material, causing irreversible changes of resistance. When large voltages have to be applied to resistors, then a chain of resistors in series should be used so that the voltage across each individual resistor is below the specified limit for that resistor type. A good rule of thumb is to make such resistor chains from 1 W resistors with a voltage drop of 500 V absolute maximum across each unit.

Manufacturing methods

The methods that are used for manufacturing resistors now almost exclusively make use of films of metal or carbon deposited over ceramic rods or tubes. Nevertheless, very large numbers of the older style of carbon composition resistors are still manufactured, and there are of course many boards still in use which contain these resistors. At one time, all resistors were either carbon composition or wire-wound.

Carbon composition resistor construction starts with a mixture of graphite, ceramic dust and resin which is pressed at high temperature and under high pressure into a rod. The rod can then be fitted with metal caps which form the attachment for the wire leads. A few uninsulated resistors were made in this form, but it is much more common to coat the rod with an insulator, ceramic or plastic. Composition resistors for high voltages are made long so as to provide a long leakage path for current; conversely, the types that were intended for higher dissipation are made of a larger diameter to provide a larger area from which to dissipate heat. Some twenty-five years ago, carbon composition resistors would be made in a wide range of dissipation values up to 5 W, but where such resistors are still available, they are mainly confined to low dissipation values, 0.125 W or 0.25 W types. Many suppliers in Europe have discontinued production of carbon composite resistors or are manufacturing them as replacement items only, but in the USA the high-quality range of carbon compositions manufactured by Allen Bradley and others are still fully available.

A typical ambient temperature range for carbon composition is −40°C to +105°C, but subject to derating above 70°C. The temperature coefficient is very high, of the order of 1200 ppm/°C, and this large temperature coefficient is one of the factors which has been responsible for the decline in this type of resistor. Another factor is the very poor stability of value. Soldering can cause a 2% change in value; the damp heat test a 15% change. One year of shelf life can cause a change of up to 5%, and 2000 hours of soak test at full rating can cause a 15% value change. These poor stability figures rule out carbon composition resistors for any application that demands reasonable reliability, and although these resistors were considered good enough for consumer electronics in the 1960s and 1970s, they are considered inappropriate for high-performance modern applications.

The fundamental problem of carbon composition resistors is that the conducting material is a mixture of conductor and non-conductor, not a pure material. The mixture conducts because grains of carbon are touching each other inside the mixture, and the non-conducting particles serve to make the path length for current longer than it would be if the resistor contained only carbon. Any mixture like this is inherently unstable, however, because the grains are simply pressed against each other, and therefore the expansion and contraction caused by heating and cooling must invariably make some difference to the contact between grains, leading to changes in value. As a result, carbon composition resistors are not now specified in any new equipment for any purpose except for their ability to withstand current surges.

The maximum working voltage for carbon composition resistors of 0.5 W or above is 500 V, but this falls to 150 V for the physically small 0.125 W types. Even the insulation resistance is poor compared to other types, around $10^9 \, \Omega$ and so about a factor of ten worse than other types generally. One of the worst features, as might be expected of a mixture of materials that relies on contact between grains of conductor, is the additional noise level that arises when current flows. This adds to the normal Johnson noise, and typical values, in $\mu V/V$, are:

R	1 kΩ	10 kΩ	100 kΩ	1 MΩ
Noise	0.03	0.62	1.3	1.8

Although this can lead to reasonably low values for the low-value re-sistors, the fact that the added noise level depends on resistance value is yet another factor that has condemned carbon composition resistors to replacement purposes only for many years and even-tually to their disappearance for many purposes.

● Another problem is the Boella effect of stray capacitance between grains, which causes the effective resistance of carbon composition resistors to decrease at VHF and higher frequencies.

The predominant method of making resistors nowadays is by using films, thick or thin, of pure conductors rather than the type of mixtures typified by carbon composition. The methods of making thin films of carbon and metals are predominantly evaporation and sputtering, although chemical deposition is also possible, and films of metal oxides are most easily deposited by chemical methods. The films are laid on substrates, usually rods or flat plates, of insulators which are almost always ceramics. The main advantages of these methods are that pure materials are used and the very nature of the deposition process ensures that a high standard of purity (lack of contamination) is maintained through the whole process.

Evaporation is most easily carried out on metals. In air, most metals will oxidize, possibly burn, at high temperatures, so that to evaporate a metal the metal must be contained in a vacuum con-tainer. When the vacuum is good enough, meaning that the pressure of the surrounding air is very low, then enough metal vapour will be present even at a comparatively low temperature just above the melting point of the metal, to coat any colder surface. In practice, tungsten wires are wound over with the metal that is to be deposited, and the substrate material is arranged so that there is an unshadowed path between the tungsten wire and the substrate. The container is evacuated, and the tungsten heated by passing several amps through it. The metal wire around the tungsten will melt and then evaporate, so that a film will condense on to the substrate material. The thickness of film can be gauged by using one dummy substrate which is fitted with electrical contacts between which the resistance can be measured during the evaporation process.

Sputtering is a more rapid process which deposits coarser and more powdery films. The material to be deposited is in the form of a plate, with another conductor present at a distance, so that a high voltage can be applied between the conductors. The container is filled with a gas such as argon, and the pressure is then reduced, but not to such a great extent as would be required for evaporation. A high voltage is applied between the plate of sputtering material and the other conductor, and at the gas pressure that is used, this sets up a gas discharge in which the plate of material is bombarded by fast ions. This bombardment displaces atoms of the material and ionizes them so that they will move with the direction of current and some inevitably will land on the substrate. If the sputtering material is a conductor, the resulting film will be a conducting film.

Metal oxides can be deposited by chemical methods, which usually involve reacting the pure metal with a gas at a low pressure and a fairly high temperature. One well-established resistive film is tin oxide, which can be created directly by heating the body of the resistor in tin chloride vapour. For other oxides, it is more usual to lay down a thin film of metal first and then react it with oxygen, measuring the extent of the conversion from metal into oxide by monitoring the resistance of a test piece.

Metals can be evaporated and sputtered, and this applies also to carbon, although the temperatures for evaporating carbon are very high and best achieved by focusing an electron beam on to a piece of carbon. Achieving conducting films, however, is only part of the process of making a film type of resistor, because the films, unless they are made perilously thin are of too high a conductivity for any but the lowest values of resistance. To achieve a high resistance value, a film must be long and with a very small area of cross-section. Films of 100 kΩ per square are now achievable, allowing resistors with values of GΩ (1 GΩ = 1000 MΩ) to be constructed.

In order to achieve the final resistance value, then, the film has a screw thread cut into it, simultaneously making it long and of much smaller area of cross-section. Such a spiral track can be very precisely controlled and can make the resistance of the final product up to a thousand times greater than the resistance of the original film on the same substrate. One of the attractions of the process is that it is possible to monitor the resistance of the resistor as the track is being cut and use this value to control the cut, so that resistors of very precise value can be obtained. Given this principle, then, the main differences between modern resistor types lie in the choice of materials and the way in which the resistor is finished. The main materials for film use are metals (in particular alloys which have comparatively high resistivity and low temperature coefficient), metal oxides, and to a lesser extent, carbon.

Carbon film resistors have replaced the older high stability (HS) type of composition resistors in the dissipation ranges of 0.25 W to 2 W. The ambient temperature range is $-40°$C to $+125°$C, subject to the usual derating above 70°C. Temperature coefficients are usually negative and in the range of -50 to -1000 ppm/°C – the higher values of resistance have the higher temperature coefficients. Minimum insulation resistance values are of the order of 10^{10} Ω normal for all types of film resistors, and the stability of value is very much better than for composition. Worst-case values are quoted as 0.5% change on soldering, 4% maximum on the damp heat test, 2% on a shelf life of one year, and 4% after 2000 hours on full ratings. The noise level is quoted as 1 μV per volt for any value of resistance, unlike the composition types. Working voltages range from 250 V for the 0.25 W sizes to 750 V for the 1 W and 2 W sizes.

- The percentage changes quoted above are for resistors not intended for precision purposes, and much better figures can be expected if you select better types.

Although carbon film resistors are a considerable improvement on composition, the use of carbon invariably leads to the rather high temperature coefficient, and because pure carbon is used, this value is negative. In addition, carbon film resistors have poorer figures for almost all of the measured features for resistors with an exception of maximum working voltage. This usually rules out carbon film for premium uses, particularly for high-reliability military uses, although a military specification for such resistors (MIL-R-11) exists.

Carbon film is preferred for consumer electronics if prices are low enough, but since metal and oxide film type are just as cheap to produce in quantity there is very little advantage in the use of carbon film, and the range steadily decreases. Carbon film resistors are still the least costly method of obtaining resistors of dissipation above 1 W combined with reasonable stability, but the demand for resistors of this type is steadily dwindling, and 2 W carbon film resistors are being phased out from the lists of most manufacturers. This leaves metal film and metal oxide films as the predominant resistor types for the 1990s.

There is no single metal film resistor type, because there is no standardization of the type of metal or thickness of film. This means that the characteristics of a 0.5 W metal film resistor may be rather different from those of a 0.25 W resistor, and these in turn can differ from the characteristics of a 0.5 W resistor which is made using a thicker ceramic–metal film (a cermet or metalglaze film), or a resistor made with high-resistivity cermet thick film. Metal film resistors must therefore be chosen with some care, particularly when a resistor from one source is to be replaced by a resistor from another source.

The ambient temperature range is generally $-55°C$ to $+125°C$, although the thicker films can accept ambient temperatures up to $+150°C$. This, however, does not apply to the thick high-resistivity film types, whose temperature rating is considerably narrower, typically $0°C$ to $+130°C$. There are also considerable variations in temperature coefficients with values that range from 50 ppm/$°C$ (0.25 W thin film) to 100 ppm/$°C$ (0.5 W thin film or thick film) and up to 300 ppm/$°C$ for the high-resistivity thick-film types. Maximum working voltage levels range from 250 V for the 0.25 W types to 1000 V for the high-resistivity thick-film types.

Noise levels are very low, of the order of 0.1 μV/V for the thicker films, but the 0.25 W types have higher noise levels of the order of 0.25 μV/V for values below 100 kΩ and around 0.5 μV/V for higher resistance values. The high-resistivity film types also have noise levels of around 0.5 μV/V. All types have high insulation resistance figures, at least 10^{10} Ω.

The superiority of metal film types over carbon once again shows in the figures for stability. The change on soldering can be expected to be less than 0.25% and for some types will be less than 0.1%. Damp heat testing causes changes of typically less than 1% and the one year shelf life test gives changes that are typically less than 1%.

More significantly, the endurance test at full rating for 2000 hours causes changes that are often less than 1%, with only the high-resistivity thick-film types showing changes as high as 2%. Improvements in manufacturing techniques now make 1% seem unacceptably high, and figures of 0.1% or less can now be achieved.

The cermet type of film resistor is often described as a *metal glaze* type because the film has many of the characteristics of glass, being semi-transparent and very hard. Cermet films are almost invariably deposited on ceramic rods or plates and are capable of withstanding high temperatures better than pure-metal films. This allows metal glaze resistors to be made in very small sizes but with a disproportionately high dissipation, typically 0.5 W dissipation for a body size that is identical to that of a 0.25 W metal film type of conventional construction. These metal glaze resistors have to be encased with materials which can withstand the temperature, and plastics of the phenolic type have been developed to suit.

Metal oxide films form a slightly different category, although many of their characteristics are similar to those of metal film types. They have in the past been easier to construct than the metal film type, and were the first type to replace carbon composition in general use, although they are found to a much lesser extent nowadays and are currently disappearing from the pages of suppliers' catalogues. The temperature coefficient of around 300 ppm/$^{\circ}$C is rather higher than will be found for metal film types, but in other respects, notably stability of value, the metal oxide types are comparable with the best metal film types.

Mention must also be made of packaged resistors, meaning a set of resistors, usually eight, which are formed on to an IC DIL package, or with a single set of inline pins. These resistors are usually of thick-film construction, and can be obtained as individuals or commoned (with one end of each resistor taken to a common pin). The resistance ranges that can be obtained are very limited; they correspond to the uses of these resistors which are mainly for 'pull-up' purposes in digital circuits so that only a few values, mainly in the range $1\,k\Omega$ to $10\,k\Omega$ will ever be used. Temperature coefficients of $0.5\,ppm/^\circ C$ and accuracy of matching of 0.01% to 0.1% can be achieved.

Wire-wound (WW) resistors form a separate group, distinguished by being the oldest form of resistor type still manufactured. As the name suggests, these resistors are constructed by winding a length of insulated wire around a former, and the resistance value can be very precise, since the resistivity, length and area of cross-section can all be precisely controlled. The main problem of construction is that a long length of wire is needed for all but the lowest values. Values of as low as 0R03 can be constructed easily, but for the higher values of several kilohms the wire needs to be of a very small cross-section and also of a considerable length. For example, if a high-resistivity alloy is used with a value of $45 \times 10^{-8}\,\Omega m$ then it is reasonable to specify a diameter of about 0.2 mm. For a resistance value of 4K7, however, this calls for a wire length of 328 m, a substantial length to try to accommodate in the space that would be acceptable for a 4K7 resistor. Nevertheless, a resistance of $100\,k\Omega$ can now be achieved in a wire-wound of only $\frac{1}{8}$ inch $\times \frac{1}{4}$ inch diameter.

For the lower dissipation ranges, very thin wire can be used, but such wire has to be handled very carefully and must be protected carefully once wound, because on damp-heat tests, any penetration of moisture can set up enough electrolytic corrosion to corrode parts of the wire away and sever contact.

The continuing demand for wire-wound resistors arises from two features that make them unique. One is the very close tolerance of resistance value that can be achieved without selection. Knowing the resistivity and area of cross-section for the wire, the length of a desired resistance value can be calculated and a tolerance achieved that depends on the uniformity of the area of cross-section. For close-tolerance wire-wounds, however, an excess length of wire is used, and is cut to length while the resistance value is being measured.

- The other feature of the wire-wound resistor is that it can be constructed for very high dissipation values, 50 W or more if need be. The construction is quite unlike that of a precision resistor.

Since the wire element is of pure metal or alloy and is thick compared to the metal film, oxidation of the surface will not necessarily cause rapid breakdown, and in any case the surface can be protected by glazing or ceramic coating.

One less desirable feature of wire-wounds, however, is thermal EMF. This is a form of contact potential that arises from the use of a resistor wire that is not the same as the wire used for connecting leads. The voltage (DC) that is generated at this pair of junctions in the resistor will alter as the temperature alters and can cause problems in measuring circuits unless it can be balanced out. Wire-wound resistors that are intended for precision uses in measuring equipment should have very low values of thermal EMF, typically $0.2\,\mu\text{V}/^\circ\text{C}$.

In addition, if the wire is wound in a spiral the resistor will have an appreciable amount of inductance, and this is undesirable. Non-inductive windings are used, reversing the direction of the winding at intervals so that the magnetic fields due to each section will cancel out. The cancellation is never perfect, and wire-wound resistors have much higher values of self-inductance than carbon or metal film types. Care needs to be taken to avoid the use of wire-wounds where their inductance might cause problems.

Wire-wound resistors fall into two very distinct classes, those made for use in measuring instruments for which precision and stability of value are the required characteristics, and those made for high dissipation. Taking the instrument type first, the wire can be of very small cross-section, but not to the extent where the tolerance of cross-section becomes too great. The required tolerance of resistance value would be of the order of 0.1% or better, and the temperature coefficient of resistance will be very low, around 3–5 ppm/$^\circ\text{C}$ as compared with the 25 ppm/$^\circ\text{C}$ for most metal film types. The stability of value is very important, and values such as 35 ppm per year for a resistor working continually at full rated power are quoted. Such resistors never run at full rated power, so that the stability is always considerably better than this figure would indicate. The thermal EMF will be very low, and it is normal for the ratings to be very conservative, so much so that the temperature rise at full rated power will be only around 30°C. Values are generally in the range $10\,\Omega$ to $10\,\text{k}\Omega$, and since these resistors are not subject to high temperatures they can be encapsulated in epoxy resin

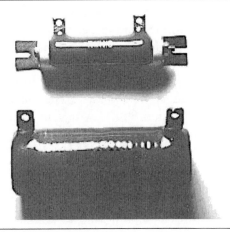

Figure 2.9 Typical high-dissipation wire-wound resistors.

materials. Typical applications are precision AF attenuators and switched gain controls, measuring bridges, test circuits, comparators, calibration equipment and strain-gauge metering circuits.

The much more extensive range of wire-wound resistors for high-dissipation uses can be divided up according to the type of coating on the resistors; most are of 5% tolerance. At the lowest dissipation levels, silicone resins can be used to provide compact resistors which can withstand high temperatures, with a temperature rise of 300°C above ambient temperature possible. A more common (and traditional) type of coating is a vitreous enamel which uses a glassy type of coating. This coating has a very good insulating resistance when cool, but its insulation resistance at full rated temperature is considerably lower, a feature that is common to many types of glass, most of which are conductive when molten. Vitreous coated resistors are nowadays less common because of this feature, because many applications of high-dissipation resistors call for maintenance of insulation resistance at working temperature, and a vitreous type cannot be regarded as an insulated resistor when it is hot. The maximum working surface temperature is around 400°C (*not* 400°C above ambient), and the typical temperature coefficient of resistance is 75 ppm/°C, although values as high as 200 ppm/°C can be encountered. Values are in the 1 Ω to 10 kΩ range.

Most of the high-dissipation wire-wound resistors that are used currently are of the ceramic type, with the wire element wound on to a ceramic core and protected by a ceramic coating. This permits good heat dissipation, high insulation resistance and good physical protection of the wire element against damage. Dissipation values are in the 4 W to 17 W range, and resistors of this type are usually fitted with leads that allow for either horizontal or vertical mounting (along with support pillars). The maximum surface temperature is usually around 300°C with temperature coefficients in the range of 250 to 400 ppm/°C. The smallest types allow for a maximum resistance value of 10 kΩ, but the larger types can be obtained with values up to 22 kΩ.

The highest dissipation values are obtained using resistors which are combined with heat-dissipating aluminium fins. These resistors are constructed using a winding on a ceramic former, coated in silicone resin and encased in an aluminium extrusion which is surface anodized in order to maintain very high insulation resistance (of the order of $10^9\,\Omega$) along with reasonable thermal conductivity of around 4–5°C per watt. Typical dissipation ratings for such resistors are 25 W to 50 W, but these ratings assume that the heat can be dissipated to other metal surfaces. If the resistor is to be used in free air, not clamped to a metal chassis, some derating will be needed. The maximum surface temperature for this construction is 200°C and temperature coefficients are low, typically 25 ppm/°C for the ohmic values above 47 Ω. The temperature coefficients are usually higher for the lower resistance values, and can be very high for resistors in the milliohm ranges.

Table 2.7 compares the characteristics of modern fixed resistor types. The values are typical values collated from a range of manufacturers and give a reasonable view of the average characteristics of resistors excluding wire-wound types.

Brief circuit notes

The uses of fixed resistors are in setting current values, using a single resistor, and in setting voltage levels using a voltage divider circuit. The application for setting current levels makes use of the $V = R \times I$ relationship, and the tolerance of current will be the same as the tolerance of resistance if the voltage is fixed. In circuits that use separate (discrete) resistors as compared to resistors incor-

Table 2.7 Comparisons between resistor types

Characteristic	Carbon comp.	Carbon film	Thick metal film	Metal film	Precision metal film
Temp. range	−40 to +105	−55 to +155	−55 to +130	−55 to +125	−55 to +155
Temp. coefft.	1200	250–1000	100	100	15
V_{max}	350–500	350–500	250	200–350	200
Noise	4 (100k)	4 (100k)	0.1	0.5	0.1
R insul'n	10 000	10 000	10 000	10 000	10 000
Solder	2%	0.5%	0.15%	0.2%	0.02%
Damp heat	15%	3.5%	1%	0.5%	0.5%
Shelf life	5%	2%	0.1%	0.1%	0.002%
Full rating	10%	4%	1%	1%	0.03%

Notes: Temperature coefficient value is maximum
Noise level is in units of μV per volt of applied DC
Insulation resistance is in MΩ
Percentage changes in value are for soldering, damp heat, one year storage and 2000 hours of continuous use at 70°C, full rating.

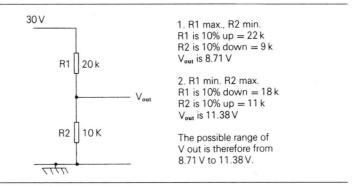

Figure 2.10 The effect of 'worst-case' tolerances on a voltage divider circuit.

porated in an IC, the tolerance of resistance can be quite close thanks to the almost universal adoption of metal film resistor types.

Tolerance of voltage levels is not quite so simple in a typical resistive divider circuit (Figure 2.10). If we suppose, for the sake of illustration, values of 10 kΩ and 20 kΩ with a 30 V supply, the ideal voltage output across the 10 kΩ resistor would be 10 V. The worst cases for tolerance will be when the two resistors are at the opposite tolerance scales, and if we suppose that the tolerances are each 10%, then we can see what effect this would have on the output voltage. With the 20 kΩ resistor 10% high and the 10 kΩ resistor

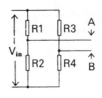

Worst-case example
R1, R4 10% up, R2, R3
10% down. Suppose that
all resistors are nominal
10 k, then R1, R4 are 11 k
and R2, R3 are 9 k. This
makes voltages:
A ..... $0.95 \times V_{in}$
B ..... $0.55 \times V_{in}$
instead of being equal.

Figure 2.11   The effects of tolerances in a bridge circuit, showing 'worst-case' assumptions.

10% low, the output voltage is 8.71 V, rounding the result to two places of decimals. This is about 13% low. With the 20 kΩ resistor 10% low and the 10 kΩ resistor 10% high the output voltage is 11.38 V, again rounding the result. This is 13.8% high.

A good rule of thumb in this type of circuit is that the overall tolerance of voltage will be worse than the tolerance of the resistors, and can be found by multiplying the individual resistor tolerance by 1.4, assuming that the resistors are of equal tolerances. In a bridge circuit (Figure 2.11), which consists of two voltage dividers, the effect of tolerances can be even greater, because one voltage may be 10% high and the other 10% low, assuming a bridge circuit made from 10% tolerance resistors. The example shows a bridge circuit in which the voltages at points A and B should be equal at 0.5 V, but are instead 0.1 V different.

- The use of 10% resistors in this example emphasizes the effect. In practice, using 1% resistors would make the effect much more acceptable. Using multiple resistors in series or parallel can also reduce the effect of tolerances.
- Another ploy is to use resistors from the same batch, so that their deviations in value are likely all to be in the same direction.

Resistance measurement can be carried out quickly and crudely, using a current-metering method, or more precisely using a bridge circuit. The method that is used for analogue multimeters is illustrated in Figure 2.12. The battery is connected in series with both a fixed and a variable resistor, and the output leads are shorted so that the meter can be set to give full-scale reading (FSD) by adjusting the variable resistor. This compensates for changes in the EMF

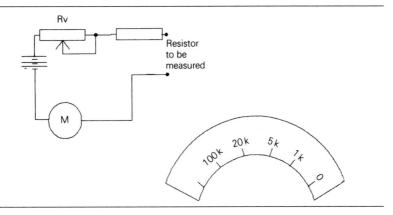

Figure 2.12   The method that analogue meters use for resistance measurement. This causes the scale of resistance to be very non-linear.

of the battery, and when this has been done, the unknown resistance value is connected between the leads, so that its resistance value can be read from the meter scale. The scale markings are reversed as compared to current and voltage scales, and are non-linear, with equal scale distances representing a few ohms at one end of the scale and many kilohms at the other end.

Digital meters make use of a constant current supply which is passed through the unknown resistor, using the meter to measure the voltage across the resistor and so display a resistance value. The considerable advantage of the digital meter in this respect is that the reading is much easier to make and there is no problem of inter-preting the reading of a pointer on a non-linear scale. The precision of measurement is usually better than 1%, a figure that cannot be approached by most analogue meters. A good digital resistance meter can provide even better measurement accuracy than a bridge measurement.

Bridge circuits can give precise measurements, but more time is needed and the precision depends on the tolerances of the resistors that are used in the bridge. If a bridge method is being used because meter methods are too imprecise, then the resistors in the bridge must be to 0.1% tolerance or better.

• Resistors are now manufactured with values in the range 0.002 Ω to 0.1 Ω, and these present a new set of problems of measurement.

Chapter 3

Variable resistors, potentiometers and diodes

When a resistor, which can be constructed by any of the processes that have been dealt with in Chapter 2, is provided with a third contact that rubs on the resistive material and makes contact with it, then the amount of resistance between this third contact and either of the fixed contacts can be made variable since it will change when the position of the contact on the resistive surface is altered. This type of arrangement can be described as a rheostat, variable resistor, trimmer or potentiometer according to its prospective use. The common factor is that the position of the third contact is set mechanically, so distinguishing this type of resistance variation from the type that is found in photoconductive cells, where the resistivity of the material is changed by light, or in thermistors in which the resistivity of the material is changed by an alteration of temperature. Devices in which the resistivity of the material itself is changed are dealt with in Chapter 7.

The three names that are used reflect the developments in the use of these components. The first radio receivers that used thermionic valves with tungsten filaments (bright-emitter valves) needed some method of controlling the filament temperature, and this was done by including a variable resistor in the circuit. Such a variable resistor or rheostat (the name means constant current) consisted of

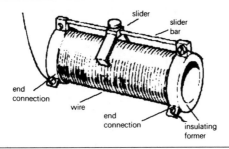

Figure 3.1 The old traditional type of variable resistor or rheostat, still used for some high-current applications.

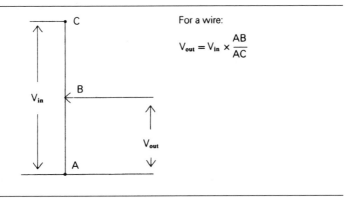

Figure 3.2 The simplest form of potentiometer, using a straight wire. The length of wire between ground contact and tap position determines the voltage output.

a wire winding on a ceramic tube, with one fixed contact and one moving contact (Figure 3.1), so that a variable length of resistance wire could be put in series in the circuit. This controlled the current by controlling the total resistance of the circuit. The term variable resistor therefore refers to this use of one fixed and one moving connection.

The term potentiometer is derived from the old established method of obtaining a precise division of a potential (a voltage), by using a resistance wire connected as indicated in Figure 3.2. The fraction of the supply voltage that is obtained at the output is proportional to the length AB of the wire if the wire is of uniform cross-sectional area. This use of a resistor requires two fixed connections and one variable connection, so that any variable resistor with three connections became known as a potentiometer.

The term trimmer, and the trademark trimpot, are of much more recent origin, and seem to have been used first for small preset capacitors. When superhet radio receivers were first constructed, the oscillator frequency was controlled by a variable capacitor which was one of a ganged set (of two or three). In order to keep the swing of frequency the same for the oscillator as for the main tuner(s), a trimmer preset capacitor was wired across the variable (in parallel) and a padder preset in series. The name of *trimmer* then started to be used for any form of preset adjustable capacitor or resistor, which would be set once and then left at its setting, being adjusted again only if some change in another component made it necessary. The essence of a trimmer variable resistor is that it should be small and not provide for frequent adjustment. Adjustment is often by means of a screwdriver, preferably an insulated screwdriver.

For electronics purposes, all forms of variables have three contacts, and when variable-resistor applications are required, one of the fixed connection points can be connected to the variable contact. The old style of variable resistor is still encountered in some applications (theatre lighting in the few places where thyristors are not yet in use), but none of these need concern us. For a three-contact potentiometer, the three important features are the angle of rotation of the shaft, the division ratio, and the law (taper). The division ratio is the ratio of the resistance from one fixed contact to the moving contact to the total resistance between the fixed contacts. The law (or taper) of the potentiometer expresses how the division ratio is related to the rotation.

In some applications, the *resolution* of a potentiometer may need to be known. The resolution is the smallest alteration in resistance that can be achieved, and is usually expressed as a percentage. For example, a resolution of 1% would imply that 100 separate steps of division ratio could be achieved, and for a 1 kΩ linear potentiometer this would imply that the change of resistance in each step was 10 Ω.

The materials that are used for the resistive tracks of potentiometers and trimmers are the same as those used for fixed resistors, so that the information contained in Chapter 2 need not be repeated. Carbon composition tracks have lingered rather longer in potentiometer production than on fixed resistors, but the predominant technology is now the cermet track. The variation that we

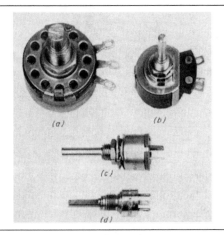

Figure 3.3 Typical potentiometer shapes.

see in potentiometers and trimmers tends to be more in the physical form than in materials or specifications, but later in this chapter we shall refer to the various potentiometer laws (tapers) in more detail.

Carbon-track potentiometers

The carbon-track potentiometer is a well-established design, and some types have been in continuous production for more than fifty years. The carbon track is moulded into an arc of a circle, covering about 315° of the circle, and connections are made by clamping metal contacts against the ends of the track. The moving connector is carried on an arm that is fixed to the rotating shaft of the potentiometer, and on the highest grade of potentiometers carries a spring-loaded carbon brush (using the word as it is used in connection with an electric motor) which makes contact with the carbon track. The carbon-to-carbon contact has fairly low friction, causes little wear of the track, and generates comparatively little electrical noise. The problem is to ensure a good connection between the moving contact and its connector on the casing. The usual solution is to make the springing in the form of a leaf, and have a separate electrical contact, using a piece of braided wire (the pigtail), Figure 3.4. The methods that are used for brushes of electric motors, using

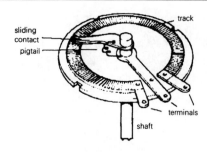

Figure 3.4 Connection between the moving contact and the terminal can be made by a sliding shoe on a copper track or by a braided wire 'pigtail' or both.

coil springs, do not give a reliable enough contact for signal purposes. As an alternative to the use of a pigtail, some designs use an arc of copper fixed to the third connector pin, and the spring which makes contact with the brush also makes connection with this metal arc. Since this scheme relies on two sliding contacts, it is more prone to contact problems than the pigtail type.

The construction of the carbon composition potentiometer ensures that the spindle is insulated from all of the potentiometer contacts. The standard spindle diameter is 6.35 mm ($\frac{1}{4}$ in) and the potentiometer is clamped in place by the spindle bush which is of 9.5 mm diameter and threaded. The standard spindle has a length of 66 mm, and the end is cut to a flat to allow knobs to be clamped in place. The track of the potentiometer is covered by a metal casing, whose overall diameter is often 30 mm or 50 mm, although other sizes are used. The mechanical motion of the spindle covers 315° of rotation, but since part of this rotation includes the end connections of the track, the effective maximum rotation for resistance variation (the electrical rotation) is 280°. The tolerance of value is 20%, and the rated dissipation depends on the type of law (see later), 1 W for logarithmic tracks, 2 W for linear tracks. These rated dissipations are for an ambient temperature of 40°C which is considerably lower than the usual 70°C which is assumed for fixed resistors. This reflects the use of potentiometers as being panel mounted and adjustable by the user, therefore subject to room temperature rather than to the temperatures inside a crowded casing. Temperature limits should be observed carefully, because operating a potentiometer of this type at or beyond its limit can result in a short life. Connections to this type of potentiometer are often the old style of wire-solder tags rather than pins that are intended to match a PCB.

Potentiometers of this type are usually offered with an optional switch, usually of the DPST (double-pole, single-throw) type. The switch is usually rated at 250 V AC, 2 A, and is intended to be used as a mains on/off switch. Domestic equipment has made use of such switched potentiometers in the past, but several objections exist. One is that the switch is not a substantial component and is not really suited, on grounds of electrical safety, to being the only method of interrupting the supply voltage on a device that is permanently plugged in. The other point is that operating the switch requires the potentiometer shaft to be turned to its extreme anticlockwise position, so that there is considerably more wear on the track than would be the case if the switch were a separate component. In general, mains-operated domestic equipment no longer uses switches attached to potentiometers, and they are produced for replacement purposes only. Resistance values for stock components range from 5 kΩ to 12 MΩ in log tracks, and 100 Ω to 1 MΩ in linear tracks.

The standard 30 mm diameter type of carbon composition potentiometer has superseded a vast range of uncoordinated sizes that formerly existed, but several smaller sizes also exist. The midget style of composition potentiometer has a body diameter of 2.3 mm, and is rated at 0.5 W linear or 0.25 W log track at the usual temperature of 30°C. These midget types have a similar amount of mechanical and electrical rotation, but if a switch is fitted, the mechanical rotation is about 350°. The most common type of connecting tags are of the old style drilled lug, but many other types are used.

The midget variety is also available as a twin-gang potentiometer, with two separate potentiometer tracks in a twin casing and driven by the same shaft. The important feature for a ganged potentiometer is the matching of the tracks, so that for any given position, the division ratio of the two potentiometers should ideally be equal. This is exceptionally difficult to achieve, and the usual specification calls for the two division ratios to match to within 2 dB, which in voltage terms means that the division ratios can differ by a factor of, at most, 1.26. This could mean that when one division ratio was 0.5 (voltage from moving contact = half of voltage across the whole potentiometer), the other division ratio was as high as 0.63 or as low as 0.4. This would be unacceptable for measuring circuits, but for low-grade audio applications (low-cost stereo amplifiers) in which these components are used, this is acceptable. Resistance ranges for single potentiometers are $5\,k\Omega$ to $2\,M\Omega$ for log tracks, $100\,\Omega$ to $1\,M\Omega$ for linear tracks. Ganged types are available in a very limited range of values only, usually up to $100\,k\Omega$ only.

The sub-miniature carbon composition potentiometer has a body diameter of only 16 mm and uses a spindle of 4 mm diameter. The connections are made through pins on a 2.5 mm grid for direct soldering to a PCB. The dissipation of 0.05 W is adequate for the type of signals or DC levels that this potentiometer is designed for, and when a switched version is specified, the switch is an SPST type rated at 12 V DC, 2 A rather than for mains supplies. The resistance range for potentiometers of this type is very restricted, with sometimes only one value being available from one manufacturer.

Slide potentiometers

The rotary form of potentiometer has been the traditional form for carbon composition tracks, but even before the days of radio construction, wire-wound variable resistors and potentiometers existed in linear form. The wire element was wound on a ceramic tube held between two end-cheeks (as in Figure 3.1), and connections could be made at each end and also to a metal rod against which a sliding contact was sprung. The linear form gives a better visual indication of the division ratio to which the potentiometer is set, and a large number of such potentiometers can be placed together, making it very obvious if one is at a different setting from the others.

Sliding carbon track potentiometers are more recent, and were introduced on domestic electronic equipment in the 1960s. By mounting such controls vertically, it was made much easier for the user to associate the position of the control with the quantity (brightness, volume, contrast, colour) that it controlled. Since switches are never included with slide potentiometers, their use started the trend to separate mains switches.

• Vertical slide potentiometer controls in the UK are arranged so that the minimum position is at the lower end; in the USA the minimum position is at the upper end.

A typical slide potentiometer uses a track length of 55 mm, with a mechanical travel of the wiper of 58 mm. Both log and linear tracks can be obtained in a resistance value range of around 10 kΩ to 100 kΩ. Ganged assemblies, with a metal screen between units to reduce cross-talk, are available, and matching bezels and knobs also exist for these units. Slide potentiometers are extensively used not only on domestic equipment but also on a large range of professional equipment, notably on audio studio mixers, faders and control consoles. Their use on graphic equalizers allows the user to see from the appearance of the settings on the control panel the shape of response that has been arranged, something that would be virtually impossible to gather from a set of rotary potentiometers which would also take up considerably more panel space.

Carbon presets

Carbon composition presets exist in several sizes, and as open or enclosed types. The open types have no track protection, and are intended for use in domestic equipment, or on boards that will operate in a clean air environment. All presets are manufactured with connectors that are spaced to fit on to a standard PCB. The older type of enclosed presets can be obtained with a knob adjustment or screwdriver adjustment, using casings of 16 mm diameter. The law (see later) is linear, something that is common to all presets of this type, and a dissipation rating of 0.25 W can be expected. A typical range of resistance values would be 47 kΩ to 1 MΩ, but these trimmers are by now manufactured mainly for replacement purposes, and they have been superseded by the miniature enclosed types.

Miniature enclosed trimmers are made in two forms, for vertical and for horizontal mounting, respectively, on to PCB tracks. The mechanism uses two wipers, one for the carbon track and one on a metal track, since the use of pigtails on such a small component would be difficult. The encasing of the track avoids the problems of dust during use, and solder flux condensation during fixing. The dissipation rating is typically 0.15 W at 45°C, and the voltage rating is 200 V. Tolerance is 20%, and the contact resistance variation is typically 5% (for the larger values) or 3 Ω for the smaller values. An angle of rotation of 220° (electrical) is used, and the static noise is of the order of 5 μV/V. Note that this is static noise, meaning the electrical noise that is developed with the contact left in place and not moving. For any potentiometer, much higher noise levels can be expected when the contact is moving, and there is no accepted standard for measuring this noise. The resistance range is from 470 Ω to 1 MΩ.

Although the enclosed trimmers are in many cases less expensive, it is more common to see open (skeleton) trimmers on boards, particularly on analogue equipment. Open trimmers are available in standard (18 mm × 20 mm) size and in miniature (10 mm × 10.5 mm) sizes, both equipped with pin connectors to fit standard PCBs. The standard sizes are available as horizontal or vertical fitting, and are rated at 0.25 W (at 40°C). The miniature size is usually available as horizontal fitting only and is rated at 0.1 W (at 40°C). The tolerance is the usual 20% for carbon composition.

Cermet potentiometers

Metal and metal oxide films, although excellent for fixed resistors, are unsuitable for use in potentiometers because the films are too fragile to withstand the rubbing from the wiping contact. The hard and glassy cermet materials, however, are obtainable in thick film form and are ideally suited for potentiometer use along with a carbon brush wiper arm. Cermets are, however, more common in multi-turn potentiometers and in trimmers than in ordinary rotary potentiometers, and are not used for slide potentiometers.

The hard nature of the cermet film allows potentiometers to be constructed which will have a long working life with none of the wear problems that plague carbon composition types. There are probably more radios scrapped because of a broken-down volume control than for any other reason, mainly because it is uneconomic to replace a volume control, and the use of cermet potentiometers in such applications would considerably extend working life if this were thought desirable. For more serious applications, the use of cermet potentiometers in place of carbon composition extends the trouble-free life of the circuit very considerably. For measuring instruments, cermet multi-turn potentiometers have replaced all but wire-wound, and can offer much better resolution than wire-wound types.

Cermet potentiometers are offered in 1 W, 2 W and 5 W dissipation sizes, contained in heavy-duty aluminium alloy casings. One design, intended for panel use, contains the cermet element within the actuating knob so that very little of the component protrudes inside the panel. Other versions are constructed in more orthodox style, but all are capable of much higher dissipation for their physical size than carbon composition types. Unlike the carbon types, the dissipation ratings of cermet potentiometers are for an ambient of 70°C (carbon composition types are rated at 40°C), and the temperature coefficient is of the order of 100 ppm/°C. For all but the 1 W size, a resistance tolerance of 10% can be expected – it is not nearly so easy to trim cermet tracks to precise resistance value for a potentiometer as it is for a fixed resistor. The electrical rotation angles vary from 210° to 270° depending on the physical design, and most varieties have contact resistance of around 2 Ω, with insulation resistance values in the range 10^9 to 10^{11} Ω. The resistance range can be large, 10 Ω to 1 MΩ, although several types are made in a more restricted range of 470 Ω to 470 kΩ. The cermet potentiometers that are intended for larger dissipations such as 5 W usually incorporate a finned heat-sink as part of the casing.

At the other end of the cermet application scale, there is a huge range of cermet trimmers, many of them using multi-turn adjustments. The simpler types closely follow the pattern of carbon composition trimmers, and are available enclosed or open, for vertical or horizontal mounting. The open types, obtainable as horizontal or vertically mounted, are much the same size as their composition counterparts, about 10 mm × 10 mm, but in cermet form such trimmers can dissipate up to 0.75 W at 40°C, 0.5 W at 70°C. Tracks are, as always for cermet types, linear, and the mountings are arranged for PCB use on 2.5 mm centres (0.1 inch). Corresponding enclosed types can be obtained which can be of the screwdriver adjusted variety, or finger-adjustable with a built-in knob. These are larger, with dimensions of 16 mm diameter for some, sides of 9.5 mm (square) for others, but with dissipation ratings of 0.5 W for the physically smaller types and 1 W for the larger ones. Some of the enclosed types use carbon brush wiper contacts, but others feature multicontact systems which ensure lower noise levels. A particular feature of several enclosed cermet trimmers is excellent sealing which allows them to withstand immersion in water (standard MIL-R-22097) without detrimental effects.

Most of the cermet trimmers that are found in modern equipment, however, are from the miniature range of $\frac{1}{4}$ in and $\frac{3}{8}$ in sizes, square or round. The $\frac{3}{8}$ in types can be obtained in horizontal or vertical fitting and are single-turn potentiometers with multicontact wipers. The dissipation rating is 0.5 W at the unusually high ambient temperature of 85°C, with 10% tolerance and a working temperature range of −55°C to +125°C. Several manufacturers construct these trimmers to the MIL-R-22097 water-immersion specification.

The sub-miniature $\frac{1}{4}$ in trimmers can be square or round, and the vertical type is usually square with the horizontal-fitting trimmer round. A dissipation rating of 0.4 W at 70°C applies to both types, with a working temperature range of −55°C to +125°C and resistance tolerance of 10%, although the tolerance is often extended to 20% for resistance values below 100 Ω. The resistance range is 10 Ω to 500 kΩ.

In general, the complete range of single-turn cermet trimmers will have low values of temperature coefficient, ranging from −125 to +200 ppm/°C. Maximum working voltages range from 200 V for the open horizontal type to 500 V for the enclosed; and for all except the $\frac{1}{4}$ in sub-miniature types the maximum variation in contact resistance is quoted at under 5%, and as low as 1% for the $\frac{3}{8}$ in range. The end resistance at the points where each end of the track connects to the pin is quoted at 2 Ω, although the open horizontal type can have an end resistance which will be at least 2 Ω and can be 1% of total resistance if this is the higher quantity. Adjustment of all these trimmers other than the type fitted with a knob is by screwdriver, and the effective electrical rotation angles range from 210° to 280°.

Multi-turn cermets

A single-turn potentiometer can be manufactured to an acceptable resolution, typically 0.1%, and has the advantage that the setting is not likely to vary, even when mechanical shock is applied. Whether you can set it to such precision, however, depends very much on how delicately the adjustment can be made and this is not easy particularly if the adjustment has to be made using a screwdriver.

The remedy is to use a form of gear mechanism to move the wiper. This will invariably reduce the ultimate resolution that can be attained, since the resolution permitted by the mechanism is lower than the cermet material makes possible. By using a gearing system such as a worm drive, the ease of precise can be considerably improved, making these multi-turn cermet trimmers ideally suited for 'fine-tuning' of any resistance value. The down side of this is that the setting is more likely to alter under mechanical shock conditions, and the resolution is determined by the mechanism, with all its drawbacks of gear backlash, rather than by the cermet track.

These multi-turn trimmers share the common characteristics of all cermet potentiometers in having linear law tracks, with temperature coefficients that can range from about -100 to $+200\,\text{ppm}/^\circ\text{C}$. Contact resistance variation is higher at $3\,\Omega$ or 3%, whichever is higher, and end contact resistance is in general in the range of $3\,\Omega$ to $5\,\Omega$, although the smallest types can have contact resistance values which are up to 3% of the overall resistance value. Working voltages range from $250\,\text{V}$ for the smallest types to $400\,\text{V}$ for the larger units, and the number of turns required to cover the whole resistance track range can be thirteen turns for the smallest trimmers to thirty turns for the largest. Most of these multi-turn trimmers incorporate a simple friction clutch mechanism which will prevent damage if any attempt is made to keep turning the adjuster after the end-stop has been reached.

The sub-miniature multi-turn trimmers are of square shape, measuring just $7\,\text{mm}$ maximum along each side and $5\,\text{mm}$ deep. The mechanism uses a worm and wheel to give thirteen turns from end to end, with a friction clutch drive. Even in this small physical size, a rating of $0.25\,\text{W}$ at 85°C is possible because of the use of a flame-retarding housing. Operating temperature limits are -55°C to $+125^\circ\text{C}$, and the tolerance of resistance value is 10%, with a range of values from $100\,\Omega$ to $1\,\text{M}\Omega$.

Moving to a larger size, $\frac{3}{8}$ in square trimmers have dimensions of approximately $10\,\text{mm} \times 10\,\text{mm} \times 6\,\text{mm}$ deep, and are rated at $0.5\,\text{W}$ at 70°C. The same type of drive mechanism, worm and wheel, is used, along with a friction clutch, and trimmers of the class can require twenty to thirty turns of the adjuster shaft for winding the wiper from one end to the other of the track. The temperature range is the usual $-55°C$ to $+125°C$ and the tolerance is 10% over a range of values from $10\,\Omega$ to $500\,k\Omega$.

Other types of cermet multi-turn trimmer use a rather different mechanism. The track is laid in a straight line rather than in the arc of a circle, and the wiper is driven directly by a worm screw, along with the usual friction clutch arrangement. Trimmers of this type are long, and the adjuster is at one end of a body which can be of $\frac{3}{4}$ or $1\frac{1}{4}$ inches in length. The adjustment range is twenty turns end to end, and the usual resistance range is $10\,\Omega$ to $500\,k\Omega$, with 10% tolerance. The smaller types are rated at 0.4 W at $70°C$ and the larger units at 1 W at $85°C$. The usual type of mounting is direct to a PCB, but panel-mounting adapters are available for trimmers that are subject to more frequent adjustment which must be made without the need to gain access to the PCB.

Conductive plastic potentiometers

A more modern development in potentiometers makes use of conductive plastics materials. Early attempts at making conductive plastics used mixtures that were modelled on the carbon composition type with the plastic heavily impregnated with carbon. It is now possible to obtain plastics materials that are not mixtures, but genuine conductors in their own right. However, when a potentiometer is stated to be a conductive plastic type the impregnated carbon variety is being used. The temperature coefficient of such potentiometers is usually the same as that of a carbon film type and the temperature range suggests a glass-fibre-resin mix rather than any of the known conductive plastics materials. The resolution of such potentiometers, however, is almost infinite, which is what would be expected of a genuinely conductive plastic. Figures for electrical noise are not generally available.

Conductive plastic potentiometers can be obtained in the usual panel form, or as specialized devices. In the normal panel-mounting form the size corresponds to the carbon composition sub-miniature size, around 16 mm width, but with a dissipation of 0.5 W at $70°C$. The temperature range is the usual $-55°C$ to

+125°C, and the tolerance is 10% for linear law types, 20% for log law. Electrical and mechanical rotations are of the order of 240° and 300° respectively. What makes these potentiometers, which cost much the same as their carbon composition counterparts, of interest is that they have specified minimum life and also very low-torque operation. The life is typically quoted as a minimum of 10^5 cycles of operation – it is very difficult to find any minimum life specification for other types of potentiometers. The torque figures are usually in the range of 2×10^3 Nm (equivalent to 20 g . cm), a figure which is closer to the torque for an energized motor or generator shaft.

One specialized application of a conductive plastic potentiometer is as a detector of shaft angle in a servo system. When the potentiometer is connected to the shaft of a motor, its output can be used as a measure of shaft position, and this can often be more precise than methods that are based on the use of inductive devices. The disadvantage as compared to the usual inductive method is that the potentiometer is not intended to be used for continuous rotation, only for a shaft whose rotation reverses at intervals and whose total angle of rotation is always within a 340° limit. A rotational life of at least 10^7 revolutions is possible, and the resolution is about 0.1%, with track linearity of about 0.5%. Dissipation can be up to 1 W at the lower running temperature of 40°C.

Wire-wound potentiometers

The wire-wound potentiometer, like the wire-wound fixed resistor, is of ancient origin, and its manufacture started well before there was any demand from the electronics industry. The introduction of cermet potentiometers and trimmers has, however, greatly reduced the demand for wire-wound types other than specialized high dissipation types which are not the concern of this book. One type of wire-wound, however, has not been displaced; the multi-turn potentiometer (not a trimmer) with very fine resolution and low temperature coefficient.

The two surviving types of single-turn panel-mounting wire-wound potentiometers are the 1 W and 3 W types, of which the 1 W type is obtainable in the resistance range of 10 Ω to 25 kΩ, and the 3 W in the range 50 Ω to 100 kΩ. The dissipation ratings of 1 W and 3 W are, however, for an ambient temperature of 20°C, which rules out the use of these potentiometers for many applications. Unlike fixed wire-wound resistors, it is impossible to protect the resistive tracks from corrosive atmospheres, and wire-wound potentiometers should never be specified for use in such conditions, nor where damp heat is a problem. Many designers regard these types as suitable for replacement purposes only.

The multi-turn form of wire-wound potentiometer is a comparatively expensive precision product, and is used where the finest possible resolution is required. A typical resolution, for example, is 0.040%, meaning that if the range of the potentiometer is 1 kΩ, then it is possible to rotate the shaft of the potentiometer so as to obtain an increase or decrease of resistance in steps of $0.00040 \times 1000 \ \Omega$, which is 0.4 Ω. In fact, the resolution that can be obtained varies with the total resistance of the potentiometer, ranging from 0.05% for the lowest values of 100 Ω to 0.013% for the 100 kΩ value.

The physical shape of the multi-turn wire-wound potentiometer can be a narrow-diameter long tube (typically 46 mm long by 13 mm diameter) or the more conventional potentiometer shape (diameter 22 mm, depth 19 mm), with gearing that allows ten or more turns of the shaft from one end of the track to the other. Temperature coefficient values of 20 ppm/°C for most values are common, although in the 1 W range, 40 and 80 ppm/°C are more common, and the lowest resistance values use wire with a temperature coefficient of 500 ppm/°C. The long tubular type uses a nickel-plated brass case, and has precious-metal contacts for the wiper arm; the shorter 3 W model uses a glass-fibre reinforced phenolic casing. The 3 W type has better linearity (0.25% as compared to 1% for the 1 W), lower end resistances (0.25% as compared to 2%) and a wider temperature range of −55°C to +125°C as compared to 20°C to +100°C. Some varieties that use a dial for readout can achieve linearity of 0.1% or better.

Potentiometer laws or tapers

The law of a potentiometer (called the potentiometer *taper* in the USA) describes the way in which the resistance between one fixed

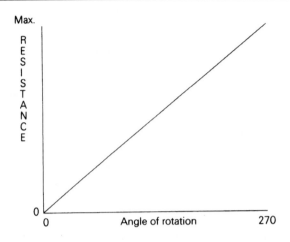

Figure 3.5 Linear law or taper, meaning that the graph of resistance from one end to the tap plotted against angle or rotation is a straight line.

contact and the moving contact changes as the shaft is rotated from the position of minimum resistance to the position of maximum resistance. The simplest law is linear, meaning that a graph of resistance (Figure 3.5) plotted against rotation angle is a straight line (usually ignoring end-resistance values). On a linear potentiometer, the change of resistance that is caused by a change of angle will be equal at any position of the shaft. For example, if a linear potentiometer has a total resistance of 15 kΩ and an electrical angle of 300° then the resistance for 30° rotation is 1K5 – one-tenth of the rotation produces a change of one-tenth of the resistance, and it does not matter whether this change is from 0° to 30°, from 80° to 110° or from 260° to 290°. The limit to how far this can be extended depends on the resolution of the potentiometer, because if the resolution of the potentiometer is poor then a small change of angle may not produce any change of resistance, or it may produce a change that is disproportionate.

Taking the example of a multi-turn wire-wound potentiometer again, a resolution of 0.040% on a 1 kΩ potentiometer means that the minimum change of resistance that can be produced is 0.4 Ω, and this on a potentiometer of ten turns (3600°) corresponds to a rotation of 0.00040 × 3600°, which is 1.44° of rotation. You can expect any well-manufactured potentiometer to permit setting to

0.1% (1 in 1000), and you should certainly not accept anything less than 0.2% (1 in 500).

The resolution of carbon composition potentiometers is considerably poorer, and resolution of cermet types is not usually quoted, so that it is not good practice to assume that linearity can be counted on to allow you to calculate what resistance change will result from small rotations of the potentiometer shaft. Linear tracks require that the track shall be made from material of uniform resistivity, and with the cross-sectional area of the track maintained constant. This is very easy to achieve with wire-wound types, reasonably easy with cermets, but rather more difficult with carbon composition, because of the variations in resistivity that are inevitable in a mixture.

Most potentiometers that are stated to be linear are not necessarily tested for linearity over their complete range, and the usual specification is that the resistance should be at its mid-value for the shaft at its mid-rotation point, with an end-resistance value of about 0.05% of total resistance. A more precisely linear response was demanded by the S-law specification, which called for checks of resistance at 25%, 50% and 75% of rotation.

The other law that is very widely used is logarithmic, or log law. In this type of law, the resistance from one fixed contact to the moving contact is proportional to the logarithm of the angle of rotation, as illustrated in Figure 3.6. This is achieved with carbon composition tracks by varying the width or the thickness (or both) of the track over the electrical angle. The purpose of using this type of law is that it corresponds to our use of a decibel scale for signal levels. The level of a signal V, compared to a standard level V_0 can be put into decibels as:

$$dB = 20 \log(V/V_0)$$

so that if the track of the potentiometer is made to follow a logarithmic law, equal amounts of rotation of the shaft should achieve equal changes of signal levels measured in decibels. This makes a potentiometer of this law well suited for controls for audio volume, brightness, contrast or any other quantity that affects the human senses, since ears and eyes also respond on a scale that is approximately logarithmic.

Describing a potentiometer as logarithmic gives little idea of how the resistance will actually vary with shaft rotation, other than that

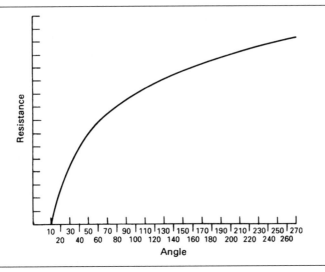

Figure 3.6   The form of a logarithmic law or taper – most of the change of resistance occurs on a small part of the angle of rotation.

this will follow a logarithmic pattern, and some manufacturers describe logarithmic potentiometers more closely as 10% or 20% types. A 10% log law potentiometer will have 10% of its resistance between the fixed contact and the moving contact for a clockwise rotation of 50% (usually with a ±3% tolerance). This implies that the resistance will rise by another 10% for half of the remaining angle of track, that is by a rotation to 75% of the full electrical angle, so that most of the resistance lies in the last portion of the track rotation. For a 20% log law, 20% of the resistance is achieved by 50% rotation, so that the rise in resistance as the rotation is continued is not so steep as for the 10% type.

If a potentiometer follows a law that is of the logarithmic form, but which has rather more than 20% of resistance for 50% travel, then over a fair part of the rotation angle the potentiometer will have an almost linear response, and over the remainder the response will be steeply logarithmic. This type of behaviour is that of a semilog law, and this type of law, along with antilog and B-law is available for specialized purposes only. It is by no means easy nowadays to find a supplier for some of these potentiometers with unusual resistance-rotation laws.

One type that is also comparatively rare now is the sine/cosine potentiometer. This uses a 360° rotation, and a continuous winding so that as the shaft rotates the voltage at the wiper contact is proportional to the applied voltage (across the fixed contacts) multiplied by the sine of the angle of rotation. This voltage will pass through zero twice in a cycle, and potentiometers of this type can be supplied with two wipers at 90° apart so that the voltage at one is proportional to the sine of the rotation angle and the voltage at the other is proportional to the cosine of the rotation angle. Potentiometers of this type have been used in the past to generate sine waves of very low frequency, but they are rare nowadays.

Notes on use of potentiometers

Wire-wound potentiometers are particularly useful for controlling DC levels, and their applications for signal voltages are limited because of the inductance of the wire-winding, so that it is unusual to find wire-wounds used for signal frequencies of more than 10 kHz. Wire-wounds, as noted earlier, are not suitable for use in corrosive atmospheres, because the wiper action usually depends on metal-to-metal contact. In corrosive atmospheres a film of oxide can appear on the surface of metals, inhibiting contact, and in moist corrosive conditions it is possible to have a voltage developed across the different metals that are in contact.

In circuits which use carbon composition potentiometers for signals it has for a long time been an article of faith that DC should not flow because of adverse effects on the noise level and possibly contact sparking. Modern carbon composition potentiometers no longer caution against combined DC and signal uses, but many designers prefer to avoid the use of carbon composition for such applications. Grease and dirt are a particular problem for carbon composition potentiometers, particularly open-track trimmers, and the rather porous nature of the composition material can make it very difficult to keep clean. One manufacturer (Bournes) maintains that reliability of a potentiometer is improved if a small current, typically 1–10 μA, flows, but opinion is divided on this point.

Diodes

Although it might seem odd to deal with diodes under the heading of variable resistors, there is some logic to it in the sense that a diode is a passive device which does not have a constant value of resistance. This is not because of any controllable variability, however, but because the diode does not obey Ohm's law. The conduction is in one direction only, and current flows in the forward direction only when the forward voltage is sufficient. A graph of forward current plotted against forward voltage is exponential, of the form shown in Figure 3.7. The threshold voltage for conduction depends on the material used for constructing the diode and for a silicon diode is about 0.6 V.

One of the main applications for diodes is in rectification, and semiconductor diodes exist in an enormous range of sizes for this purpose. The standard forms of rectifier circuits are noted in Chapter 5 (under the heading of transformers) and most make use of four diodes arranged in a bridge format. This arrangement is so common that diodes are obtainable packaged into bridge circuits, with two AC input leads and two DC output leads. The specified ratings for such diode bridges are of maximum current and RMS voltage, but the ratings are for a resistive load, so that when the

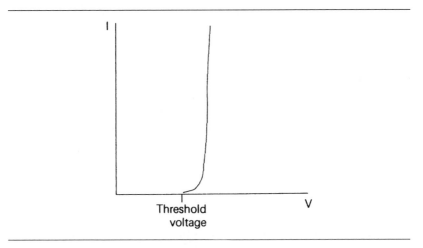

Figure 3.7 A typical diode forward characteristic. The current at levels below the threshold bias level is negligible, following an exponential characteristic so that the slope of the graph increases as the voltage bias increases.

much more usual capacitive load is used the current ratings should be reduced to 80% of the values shown. In addition, the voltage rating that is usually shown for these bridge rectifier assemblies is maximum reverse voltage. This is equal to the peak DC output voltage plus the peak AC input voltage, so that the RMS voltage equivalent is about 0.36 of this value (the reverse value divided by 2.8).

The range of these bridge assemblies for electronics use covers units up to 1200 V reverse voltage (about 425 V RMS) and 60 A (48 A into a capacitive load) of current. These high-current rectifier bridges will normally be mounted on to a heat-sink, or can be forced-air cooled for higher current ratings. Three-phase versions (using six diodes) can also be obtained in the largest current ratings.

High-voltage and high-current single diodes are also available for (mainly) application as rectifiers. These are mainly miniature types, so that the high-voltage rating is of up to 1600 V reverse rather than the high kilovolt ratings that can be obtained from series assemblies. The high-current types are also miniature 6 A rated diodes with a diameter of about 10 mm. For considerably higher current ratings, stud mounted diodes in the range 16 A to 240 A can be obtained, with peak reverse voltage ratings from 1000 V to 1200 V. EHT diodes can be constructed for a peak reverse voltage of 24 kV, and these are widely used in high-voltage supplies that make use of voltage multiplier circuitry. Good insulation is needed around these diodes because the small (12.5 mm long) size of the casing makes sparking-over very likely if dust or moisture settles on the body.

Schottky rectifier diodes are now available and have substantial advantages. A Schottky diode uses a metal–semiconductor junction and it conducts, unlike the conventional junction diode, by the movement of majority carriers. The initial application of Schottky diodes was in digital ICs, because the use of majority carriers avoids any transient conduction when the voltage is reversed from forward to backward. Another outstanding advantage is that the forward voltage drop of a Schottky diode is lower at the same current conditions than that of a comparable silicon junction diode. These features make Schottky diodes particularly useful for rectification for high-frequency waves or otherwise rapidly changing waveforms. These diodes can be obtained typically in the range of 1 A 30 V to 300 A 60 V, with the voltage rating referring as usual to reverse voltage. Schottky rectifiers are a good replacement for old germanium diodes, with the advantage that the Schottky type have much lower leakage currents.

Ultra-fast silicon diodes can also be used for switching circuits, with a typical recovery time of 2 ns or less. The recovery time is the time needed to stop conducting after a specified forward voltage has been switched into reverse. Fast recovery is required also for rectifier use on high-frequency supplies.

General-purpose diodes are used for all the range of actions other than rectification and the demodulation of signals in the higher-frequency ranges. These actions include clamping and clipping, signal shaping, biasing and demodulation for the lower frequencies. The 1N914/1N4148 series of diodes can be turned to any of these tasks. These are sub-miniature whiskerless diodes with a 75 V maximum or peak reverse voltage rating, and a reverse recovery time of less than 4 ns.

Signal diodes for use in demodulators at one time would mean germanium diodes, used because of their low forward conduction voltage. Germanium diodes are now considered an endangered species, and the diodes for use in demodulation, UHF mixing and high-speed low-power switching are all silicon based, using normal planar junctions or Schottky techniques.

Zener diodes

The Zener diode or avalanche diode is a diode that is intended to be used with reverse bias. The normal reverse characteristic of any silicon diode is to withstand reverse voltages until a threshold value at which reverse current increases rapidly. The Zener type of diode is constructed so that the breakdown of resistance for reverse voltage is particularly rapid, leading to a characteristic for the type illustrated in Figure 3.8, with a very steep slope at the breakdown voltage. This implies that over a large range of reverse currents, the reverse voltage will be almost constant, so that this type of diode is used as a constant voltage source.

A typical circuit is illustrated in Figure 3.9. The Zener diode is connected in series with a resistor which is used to limit the current. If the supply voltage input varies, but does not fall as low

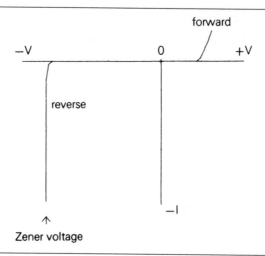

Figure 3.8 A typical Zener diode characteristic. The important portion is the reverse breakdown voltage which ensures that the voltage across the diode in the reverse direction is almost constant despite large variations of current.

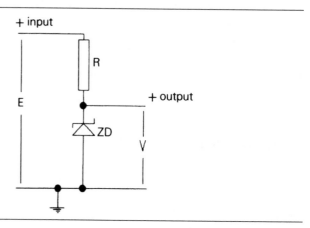

Figure 3.9 The simplest application circuit for a Zener diode, making use of the diode to provide a stable voltage despite variations in input voltage or load current, provided that the current through the diode does not drop to a low value.

as the Zener voltage, then the current through the diode will vary but the voltage across the diode will be almost constant. The current through the resistor is given by:

$$\frac{E - V}{R}$$

where E is the supply voltage and V is the Zener breakdown voltage.

The breakdown voltage is temperature dependent, but for around the voltage breakdown level of -4 V the temperature coefficient is almost zero. Below this level, the temperature coefficient is negative, and above this level the temperature coefficient is positive. This is because the observed temperature coefficient is the sum of two separate temperature effects which act in opposite directions and cancel each other out for a breakdown voltage of around -4 V.

● The graph of temperature coefficient plotted against Zener voltage does not show a rapid change around the minimum, so that many users prefer to use the 6V2 zeners as a good compromise between low temperature coefficient and low slope resistance (see later).

Another variable effect is slope resistance, meaning the rate of change of voltage as current is changed. This ideally would be zero, and in practice is a low resistance value which typically has a minimum for Zener diodes of about 7V5 rating. For most of the low-voltage diodes up to 15 V, the slope resistance is 20 Ω or less, but it is considerably greater, more than 100 Ω, for diodes in the higher-voltage regions.

Zener diodes are supplied in a range of breakdown voltages, typically 2V7 to 62 V, and in a large variety of power ratings. Small Zeners are rated at around 1.3 W, subject to a derating of typically 9 mW per degree above a temperature of 2.5 °C. Zeners of 20 W rated dissipation are available, but there is usually little point in using large wattage Zener diodes in simple stabilizer circuits, because voltage stabilization can be accomplished more cheaply and efficiently at the same power levels by using a stabilizer IC.

The more specialized forms of Zener diodes have been developed for very small temperature coefficient or for very stable voltage control. Temperature-compensated diodes can achieve a temperature coefficient as low as 0.001% per °C, and voltage reference diodes can supply a precise voltage which is very stable against temperature or current changes, and also with a very low noise output. The most precise and stable voltage reference diodes, however, are bandgap diodes which make use of the forward voltage across junctions to give a low-voltage reference source of 1.2 V.

- One form of Zener (more correctly, avalanche) diode has been developed for use as a transient suppressor, and the trademark of Transzorb is used for the General Semiconductor version. Turn-on times of less than 1 ns are quoted, along with the ability to absorb high-energy transients.

Chapter 4

Capacitors

A capacitor is an arrangement of conductors that are insulated from each other so that charge can be stored. Imagine two metal plates arranged parallel to each other and separated by air. If we apply a voltage to the plates by connecting them to a battery (Figure 4.1), then one plate will gain electrons and the other will lose electrons. When the battery connections are removed there is no connection between the plates that would allow the electrons to return, and there will still be a voltage between the plates, caused by the fact that one plate is negatively charged and the other positively charged. If the plates are now connected, a transient current will flow until there is no surplus or deficit of electrons on either plate.

The total amount of charge that flows when the plates are connected in this imaginary experiment can be calculated by multiplying the amount of current that flows by the time for which it flows, but this is not a simple calculation because the current is not constant, it follows the pattern shown earlier in Figure 1.17. A more useful way of calculating the charge emerges from the knowledge that the quantity:

$$\frac{\text{charge on one plate}}{\text{voltage between plates}}$$

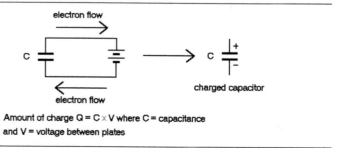

Figure 4.1 Two conducting plates separated by an insulator constitute a capacitor, which is charged when connected to a battery.

is always a constant for a particular size and arrangement of plates. This quantity is called the *capacitance* of the plates, and we can write the equation as:

$$\text{charge} = \text{capacitance} \times \text{voltage}$$

or in symbols as:

$$Q = C \times V \text{ which can also be written as } C = Q/V$$

What makes this useful is that the quantity we call capacitance can be calculated from factors such as the size and position of the plates, so that if capacitance can be calculated and voltage measured, then the amount of stored charge Q can be calculated.

When modern electrical units were being devised, the unit of charge was, as now, the coulomb, and the unit of voltage is the volt. When coulombs are divided by volts to find a unit of capacitance, this unit, called the *farad*, turns out to be too large for most of our practical purposes, although capacitors of 1 F can be and are manufactured for specialized purposes. The usual submultiples that are used for the farad are the microfarad (μF), equal to 10^{-6} F, the nanofarad (nF), equal to 10^{-9} F, and the picofarad (pF), equal to 10^{-12} F. For an ever-increasing number of purposes we also need to use the femtofarad (fF), equal to 10^{-15} F.

Capacitor behaviour

When DC or pulse waveforms are applied to capacitors, the charge and discharge behaviour is important. In any practical circuit, a capacitor will be charged or discharged through a resistor, and the time that is required for charge or discharge is expressed by the time constant, units seconds, equal to resistance in ohms multiplied by capacitance in farads, as noted in Chapter 1. A more practical set of units uses resistance in kilohms and capacitance in nanofarads to get time in microseconds, so that $1\,k\Omega \times 1\,nF = 1\,\mu s$ time constant.

- The practical interpretation of time constant, see earlier, is that the charging or discharging of a capacitor through a resistor is reasonably complete in a time equal to four time constants. For example, if a time constant is given as $50\,\mu s$, then for this CR connection, charging or discharging would be completed, for all practical purposes, in $200\,\mu s$.

For the smaller capacitors connected to any practicable resistor size, this is a short time, but for the larger types the time constant that can be obtained along with a resistor can be quite long. For example, it is possible to use a capacitor as a form of backup against short interruptions of power supply for a low-consumption circuit. If, for example, a capacitor of 1 F, 3 V working, is wired in parallel with an IC which requires a current of $10\,\mu A$ at 3 V, then this IC constitutes a load of $300\,k\Omega$. The time constant for this circuit is 300 000 seconds, more than 83 hours, and if we assume that the circuit would still operate at a lower voltage (after 63% of the voltage has been dropped) then this would constitute an 83 hour backup. In practice, the leakage current from such a capacitor would be larger than the amount indicated, and such a fall in voltage would not permit the circuit to continue working, but it is possible to maintain a useful voltage level for a drain of 1 mA for periods of an hour or more.

- An important safety point is that the larger capacitors can store a considerable amount of charge at a voltage level which can cause severe shock or burning if the terminals are touched. Even at low voltages, the amount of energy stored by a capacitor can cause destructive sparking, sufficient to melt metal wires, if a charged capacitor is short circuited.

The amount of charged energy is found from:

$$\text{Energy} = \tfrac{1}{2}CV^2$$

which gives energy in joules when the units of capacitance and voltage are the farad and the volt respectively. For example, a 5000 μF capacitor charged to 400 V will carry a stored energy of:

$$250 \times 10^{-6} \times (400)^2 = 40\,\text{J}$$

This is the amount of energy that would be achieved by a power of 40 W acting for 1 s. When a capacitor is short circuited, however, this energy can be discharged in a fraction of a second, and any brief release of such a substantial amount of energy can be very destructive. It can also be painful if it discharges across you, and possibly fatal if you take the discharge of a large capacitor through a path that crosses the heart.

When the voltage across a capacitor is continually varied, the capacitor will be continually charged and discharged, so that there will be a flow of electrons to and from the plates. This flow constitutes a current so that for an applied (sine-wave) alternating voltage there will be an alternating current flowing, and the current will be proportional to the voltage, just as the current through a resistor is proportional to the voltage across it. We can define a quantity called *reactance* which is analogous to resistance in the formula:

$$V = X \times I$$

Where V and I are AC values. However, the similarity cannot be taken too far. The reactance of a capacitor is not constant, and there is a phase difference of 90° between current and voltage (see Chapter 1). We could, in fact, draw up an alternative definition of a capacitor as an electronic device that permits the flow of signal current but not DC, and creates a 90° phase difference between voltage and current, with current leading voltage.

- When a capacitor is charging or discharging, the current *at any instant* is proportional to the rate of change of voltage. This means that instantaneous charging or discharging is impossible, since this would require an infinite rate of change of voltage.

Capacitor construction

The simplest type of capacitor is the parallel-plate type, using air as its insulation between the plates. If any solid or liquid insulator is placed between the plates, the capacitance of the arrangement is

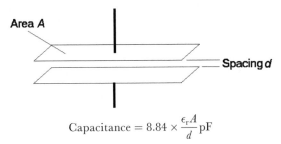

Area A

Spacing d

$$\text{Capacitance} = 8.84 \times \frac{\epsilon_r A}{d} \, \text{pF}$$

with A measured in square metres and d in metres. If units of square mm for area and mm for separation are used, the capacitance will be in nF units. The quantity ϵ_r is the relative permittivity of the material between the plates. For air this can be taken as 1.0.

Figure 4.2   The formula for capacitance of a parallel-plate capacitor. The quantity ϵ_r, is the relative permittivity of whatever insulating material is placed between the plates.

increased, and the factor by which the capacitance is increased is called the relative permittivity of the material between the plates. Figure 4.2 shows the formula which can be used to calculate the capacitance of this arrangement. This uses a quantity called the *permittivity of free space* which is a universal constant.

Capacitors for low values, a few pF, can be constructed in parallel plate form, although the use of air spacing is unusual except for special-purpose units used for radio transmitters or for calibration purposes. The way that the capacitor behaves, as well as the value of capacitance, is then considerably affected by the material, called the dielectric which is placed between the plates. The area of plate which would be required to make a capacitor of more than a few microfarads in this form is impractical, and all the other shapes of capacitors that are familiar in electronics represent methods of reducing the physical size of the capacitor and yet increasing the amount of capacitance. This process is assisted by the fact that decreasing the distance between the plates (or other conductors) increases the amount of capacitance between them. At the same time, the amount of space that would be needed to accommodate plates of large area can be greatly reduced by folding the plates or by rolling them into a cylinder. Whatever method of construction is adopted, it is always much more difficult to achieve close tolerances with capacitors than with resistors.

Table 4.1 Relative permittivity values for various selected materials. The values of relative permittivity are seldom large

Material	Value of ϵ_r
Vacuum	1.000000
Air	1.00004
Aluminium oxide	8.8
Araldite resin	3.7
Bakelite	4.6
Barium titanate	600–1200
Magnesium silicate	5.6
Nylon	3.1
Polystyrene	2.5
Polythene	2.3
PTFE (Teflon)	2.1
Porcelain	5.0
Quartz	3.8
Titanium dioxide	100

We shall return to the physical shapes and construction of capacitors later, but we need to look further at the basic design and how the use of a dielectric affects the behaviour of a capacitor. The obvious effect for which the dielectric is chosen is its value of relative permittivity, which will assist considerably in achieving the value of capacitance that is required. By its definition, relative permittivity must be a positive quantity greater than unity, and values for some solid materials are shown in Table 4.1. Most insulators have values of relative permittivity which are in the range 2 to 5, but there are several types of ceramics which have extraordinarily high values of 100 or more. This makes it possible to use these materials to manufacture capacitors of very compact dimensions, but the high value of permittivity is obtained at the expense of other features. Liquids, not shown in the table, can have even higher permittivity values; an example is Thiokol with a value of over 2000.

- Thiokol is a health hazard – you must not open a capacitor that contains Thiokol, and care is needed to avoid any error such as an overload that would burst the capacitor open.

One of the most obvious of the other features of the dielectric material is that it constitutes the insulation between the plates of the capacitor. The insulating properties of the dielectric will control the amount of current (which must be very small) which will leak between the plates when the capacitor is charged, and will also determine the dielectric strength in terms of resistance to sparking between the plates. Sparking inside a capacitor can cause damage, and although some types, notably metallized polyesters, are self-healing to a remarkable extent, others will be totally destroyed by sparking.

The leakage of the capacitor, in terms of DC leakage current, is determined by the resistivity of the dielectric. Since the dielectric is always used in the form of a large area and a thin film, only materials of very high resistivity are suitable. Suppose, for example, that we were using a dielectric of relative permittivity 5 and thickness 0.02 mm to make a 0.01 μF capacitor. From the capacitor formula (Figure 4.2), we can calculate the area of the dielectric as 4.5×10^{-3} square metres. Now if the material that we use as a dielectric also has a resistivity of 10^7 Ωm, which would be a respectable value for a material used for most other purposes in electronics, then the resistance of the dielectric film would be about 45 kΩ. This is unacceptably low for such a capacitor, and we would normally expect a value of resistance of the order of 1000 MΩ. This points to the dielectric requiring to have a resistivity which is at least 22 000 times greater than that of this example. We would not normally consider a value of resistivity of less than 10^{11} Ωm as being suitable, and for the smaller value of capacitance even this would not be acceptable. Values of the order of 10^{14} to 10^{17} Ωm can be obtained using modern plastics and also the ceramics materials that have been used for these purposes for a considerable time. High-quality capacitors can display a dielectric resistivity of 10^{18} Ωm.

DC resistance, however, is only one aspect of capacitor efficiency, one which determines for how long a capacitor can retain a charge. If a dielectric contains any charged particles, or if its atoms can have electrons dragged away by the electric field that exists across the dielectric, then when an AC signal is applied across the dielectric there will be some dissipation. In a perfect dielectric, this movement of particles would be frictionless; whatever energy was needed to move the particles out of position would be returned when they moved back. Most practical dielectric materials are surprisingly close to perfection, but for some purposes we need to know a dissipation factor, which is defined as:

$$\frac{\text{energy dissipated in one cycle}}{\text{energy stored in one cycle}}$$

and which for all practical purposes is equal to the power factor for the dielectric, the cosine of the phase angle, φ. The dielectrics that are in common use have dissipation factors which are in the range 0.01 to 0.001.

A further complication is that both relative permittivity and dissipation factor will vary with the frequency of the signal that is applied to the capacitor. The relative permittivity value will decrease as the frequency is increased, and for many materials the decrease is quite small, but suddenly increases at a very high frequency. Values of relative permittivity are usually taken at frequencies of 50 Hz, 1 kHz, 1 MHz, 100 MHz, 3 GHz and 25 GHz, although only a few materials are suitable for use at 25 GHz. Several materials exist in which the change of relative permittivity is both small and continuous, with no sudden changes.

The change of dissipation factor as frequency is changed is quite different. The change is greater, and for several types of materials it has a peak around one frequency, a resonance frequency. The relative permittivity will also exhibit a sudden change around this frequency, and the frequency at which this peak occurs is considerably affected by the ambient temperature. For a few materials, the dissipation factor is highest for low frequencies because the materials contain ions which are slow moving and which do not contribute to dissipation when the voltage across the material reverses at a high frequency. An ideal dielectric would have no peaks or other discontinuities in a graph of dissipation factor plotted against frequency.

The other aspect of the insulation of a dielectric is the dielectric strength, which is expressed in terms of the maximum electric field that can be applied to the material. The unit of electric field is the volt per metre, but for the large fields that exist across dielectric materials, units of kilovolts per metre are more suitable. Old tables show this in terms of volts per mil (thousandth of an inch) or as volts per millimetre. Figure 4.4 shows the conversion factors and a few examples of dielectric strength in terms of volts per mil and kilovolts per metre units. The range of dielectric strength is from 19 500 kV/m to 200 000 kV/m for typical dielectrics, and this figure

Table 4.2 Converting between units of electric field as volts per mil and kilovolts per metre

From	To	Multiply by
V/m	V/mil	0.0000254
kV/m	V/mil	0.0254
V/mm	V/mil	0.0254
V/mil	V/m	39 370
V/mil	kV/m	39.37
V/mil	V/mm	39.37

Example: Mica: 4000 V/mil = 157 480 kV/m or V/mm

determines the maximum working voltage (or the peak voltage) that can safely be applied to the capacitor. In general, the working voltage will be considerably lower than the value which would subject the dielectric to its rated dielectric strength.

The effect of temperature on a dielectric material will be to alter the physical dimensions, alter the resistivity and the dielectric strength, alter the relative permittivity and alter the dissipation factor and the frequency of the peak of dissipation factor (if there is a noticeable peak). The effects that alter capacitance can be measured by a temperature coefficient of capacitance which is defined in the same way as the temperature coefficient of resistance as:

$$\frac{\text{change of capacitance}}{\text{nominal capacitance} \times \text{change of temperature}}$$

and will be quoted in parts per million per degree Celsius. This temperature coefficient can be positive or negative, and negative values are fairly common. A few dielectrics can cause their capacitors to have very large negative values of temperature coefficient. Some capacitor types have such large values of temperature coefficients that they are quoted as percentage change per degree Celsius rather than parts per million. Polystyrene capacitors, now obsolete, suffer from excessive drift at temperatures above 75°C.

Finally, another effect of the dielectric is remanent charge or soakage, as mentioned earlier and dealt with in more detail in the following pages.

Quoted parameters

The quantities that are quoted for the performance of a capacitor are related to the method of construction and to the choice land methods of preparation of the dielectric, but it is very unusual to see quantities such as dielectric strength quoted. The capacitance value, maximum rated voltage and tolerance of capacitance value are the most important parameters, and tolerance values are invariably higher for capacitors than for resistors. Tolerances of 10% and 20% are normal, with close-tolerance components of 5% to 2% frequently used, and tolerances of 1% and 0.5% available for special purposes. The range of values is almost as vast as that for resistors, ranging from a few picofarads to 3F3 or more, a range of around 10^{12} from lowest to highest.

The other parameters that are quoted depend on the type of capacitor, because electrolytic capacitors (see later) cannot be judged by the same criteria as other types, and for the moment we will concentrate on the non-electrolytic types. After value and tolerance, the most important factor is working voltage, and in this respect it is usual to quote both a DC and an AC working voltage – the AC voltage rating is always lower because quoted AC voltages are RMS and it is the peak voltage that affects the stress on the dielectric of a capacitor.

These voltage ratings have a very important effect on reliability. Taking DC ratings first, the rating is quoted for a working temperature which is usually in the range 70°C to 85°C, and substantial derating is needed if the capacitor has to be used at temperatures outside this working range. The maximum rated voltage is selected so as to provide a reasonable working life, and the figure of acceptability is usually that of one failure per 10^5 working hours. If the actual working voltage is significantly lower than the rated value, the reliability of the capacitor may increase considerably because there is little likelihood of a transient overload exceeding the maximum permitted value. You should not, however, take such an improvement for granted.

The value of working voltage that is chosen is well below the voltage value that would represent the dielectric being subjected to the maximum allowable field (dielectric strength value), and some capacitors can withstand brief voltage peaks that are considerably higher than the maximum continuous working voltage. Only these capacitors that are rated to accept such peaks should be subjected to peak voltages, however, because the effect on the life of any other types of capacitors can be unpredictable.

AC ratings of working voltage are considerably more complicated than the DC ratings. There is a common belief that any capacitor that is rated for 240 V AC can be safely wired across the supply mains, but this is quite definitely not so, and only capacitors which are specifically stated as being suitable for this purpose (mains filtering capacitors) should ever be used on mains supplies. Quite apart from any other consideration, a capacitor wired from mains live to mains ground constitutes a leakage to ground, and can cause sensitive ground-leakage contact breakers (ELCB) to open. They also constitute a hazard to operators if there is a fault in ground wiring, since the local ground (metal cabinets, for example) will be live through the capacitor connection. For this reason, the capacitors that are used for mains filtering are of low capacitance value and are of a construction that permits continuous application of power voltage levels and frequency. Where larger values are used, a self-healing type of dielectric will be used so that any breakdown of the capacitor will be momentary only and will not lead to a low-resistance type of failure.

The AC voltage rating of a capacitor is determined by a number of factors. For low frequencies, the voltage rating must be low enough to avoid failure of the dielectric, so that the value has to be considerably lower than would be indicated from the dielectric strength of the insulator. At the higher frequencies, the dissipation loss becomes more important, because it can cause the temperature of the capacitor to rise to a level at which the voltage ratings are no longer applicable. Any capacitor which is to be used for RF voltages at low impedances (so that current could be high) or for pulses with a high repetition rate will have to be derated for working voltage, and manufacturer's advice should be sought unless a capacitor type specifically states its suitability for such purposes.

In many cases, the specification will provide for use with pulses whose peak amplitude does not exceed the DC rating of the capacitor, but with a stated limit on the rate-of-change of voltage. This might be, typically, 5 V/µs for a subminiature polyester type (see later), which will restrict its use to applications in which the rates of rise or fall of voltage across the capacitor are comparatively low.

Such capacitors could be used, for example, to couple fast-changing signals (because the voltage across the coupling capacitor is small), but not to decouple such waveforms. Care needs to be taken over specifying such capacitors in digital circuitry where fast pulses were present.

A perfect capacitor would dissipate no energy, since it has no resistance. Real capacitors come remarkably close to perfection, nearer than any other electronic component, but there is always a measurable amount of dissipation. This can be shown in an equivalent circuit as a series resistance, usually of a very small value. The value of this equivalent series resistance is not constant, and for most capacitor types its value will increase as the operating frequency is increased. The value of equivalent series resistor (ESR) is quoted only for electrolytics, for which the resistance is certainly not negligible. In addition, the ESR value is frequency sensitive.

Where any loss factor is quoted for other capacitor types, it will usually be as a power factor or dissipation factor (usually 0.001% or less), meaning the fraction of the volt-amperes (AC volts multiplied by AC amperes) across the capacitor which represent real watts.

Temperature range and temperature coefficients

The temperature range for a capacitor is determined entirely by the type of dielectric that is used. Many capacitors make use of plastics materials, so that the temperature range must be such that the material is used well below its melting point. In addition, several types of plastics materials undergo serious changes in form at very low temperatures, becoming brittle and even cracking.

The temperature coefficients of capacitors can be negative (which is more common) or positive, and a few types can have very large temperature coefficients, although values of $\pm200\,\text{ppm}/°\text{C}$ are usual. For some applications of capacitors, such as decoupling, temperature coefficients are almost irrelevant and tolerances are not particularly important either. The most critical applications are for capacitors which form part of a tuning circuit, where a large temperature coefficient could cause mistuning to occur as the temperature changed. Modern oscillator circuits would normally use

some form of automatic control, but careful selection of tuning capacitors is also necessary to prevent the tuning from falling out of the
range of the automatic controls because of variations in capacitance
of a fixed capacitor.

Voltage remanence (soakage)

An effect which has been known for many years, but which has not
always been of great importance, is that of voltage remanence, also
known as dielectric absorption or 'soakage'. This is an effect which
could be demonstrated quite well on the old-fashioned Leyden jar
type of capacitor. When such a capacitor is fully charged, it can be
discharged by a spark-over, but after a short time, another spark discharge is possible, and sometimes a third. The explanation is that
the energy of a charged capacitor exists as a strain in the dielectric
– literally a mechanical strain because electrons and nuclei are
being pulled out of their normal positions.

The short circuiting of a capacitor allows these particles to return,
but in some types of material the return takes longer than the time
of a short circuit, and the voltage across the capacitor builds up
again if the capacitor is left open circuit. Some modern types,
notably polystyrene, NP0 ceramics and polypropylene dielectric
types, have almost negligible soakage, and can be specified for uses,
such as sample-and-hold circuits, in which this factor is of vital importance. The old oil-filled paper type, by contrast, suffers from
this effect and must be handled with care.

One effect of this is that any capacitor which is used to smooth a
high-voltage supply should be handled with extreme care and
should preferably be kept short circuited all the time it has to be
handled. Another aspect is the selection of the auto-zero capacitor
on digital voltmeters. This capacitor is shorted at intervals to zero
the voltage input to the counting circuit, and a voltage remanence
can cause the zero point to be incorrect, with the remanent voltage
being added to the measured voltage. Careful selection of the
capacitor is needed for such circuits, and for any circuits in which a
capacitor may be shorted and then any voltage across its plates
measured. Manufacturers of capacitors will advise on suitable
types, but the remanence is seldom mentioned in specifications.

Capacitors which use Mylar dielectric are likely to have higher values of remanent voltage than capacitors which use traditional mica, ceramic, or paper.

For a full discussion of this problem, along with descriptions of test circuits, see the website:

http://www.national.com/rap/Application/0,1570,28,00.html

Voltage-variable capacitance

Any semiconductor junction will contain a depletion region of non-conducting semiconductor between the two conducting regions. Because of the structure of a semiconductor junction, the effective width of this depletion layer is not constant; it increases as the reverse bias on the diode is increased, so that the capacitance across the diode decreases as the bias is increased. Diodes that have been prepared so as deliberately to exploit this effect are known as varicap or varactor diodes.

Varicap diodes are classed as diodes rather than as capacitors, but since their uses are invariably as capacitors it seems reasonable to include them here. In a typical circuit (Figure 4.3), the varicap diode is made part of a tuned circuit, and is connected in series with an isolating capacitor so that the DC bias voltage can be applied through a large-value resistor. The voltage that is used to control the capacitance of the diode can be supplied direct from a

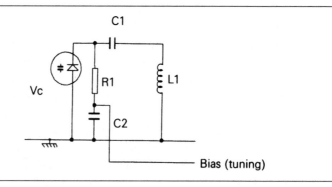

Figure 4.3   Using a varicap diode in a tuning circuit.

potentiometer for hand-tuning (eliminating the use of a variable capacitor of the mechanical type) or from a phase-sensitive detector circuit whose DC output will be proportional to the deviation from a set frequency. This latter use is the familiar AFC type of circuit.

It can be surprisingly difficult to obtain specified types of varicap diodes, because they do not appear in many components catalogues, and are often supplied by semiconductor manufacturers to special order only. The main reason is that these are seldom components that are replaceable in servicing work, and in the applications such as TV tuners, the varicap diode forms part of a circuit whose mechanical layout is critical and which is usually serviced by complete replacement of the module. In some circuits, ordinary silicon diodes can be used in place of specified varicaps.

- A varicap diode can also be used as an active component in a parametric amplifier circuit. Details are beyond the scope of this book, but briefly, amplification is achieved by altering the diode capacitance in synchronism with the applied waveform. The effect is analogous to the way that a child on a swing can increase the amplitude of swinging by making synchronous leg movements.

Capacitor types

CERAMICS AND MICAS

The names that are used for types of capacitors are the names of the dielectric materials, because the performance of a capacitor is so closely tied to the type of material that is used for its dielectric. Ceramic covers any of the materials which consist basically of metal oxides that have been fused at very high temperatures; typical raw materials are alumina (aluminium oxide) and titanium oxide. Mica is a natural material which splits into plates that can be remarkably thin; its main form is the mineral muscovite, or ruby mica. When this material is split into plates, the plates often have a silvery appearance (caused by an air film between the remaining plates) so that these are called *silver-mica*. This has caused considerable confusion because the coating of mica sheets with silver creates a composite called *silvered mica*.

Because of the natural shape of the raw material, mica is used to make capacitors which are of plate shape, circular or rectangular. Ceramics can be formed to any suitable shape, including plates and tubes, so that the range of capacitor shapes is greater for ceramics than for micas. Whichever of these two insulator types is used, the method of forming the capacitor is to deposit a metal layer on each side of the dielectric. This is easiest when the material is in plate form, and the deposition of metal can be carried out by chemical methods (a traditional method which is particularly easy for depositing silver) or by evaporation or sputtering. The metal layer has to be kept clear of the edges or wiped from the edges so as to avoid short circuits or potential spark-over points. Connecting wires can then be soldered to the metal layer, and the whole capacitor covered with an insulator which can be plastic or another ceramic material.

Tubular ceramics are formed in the same way as plate types, but the metallization process is considerably more difficult and only a chemical method can be used to put a coating inside the tube. Connection to this coating is also more difficult, but the small volume of the tubular type can sometimes be an advantage so that this type of capacitor has been used for many decades, although it has now disappeared from many catalogues because it can be made only in the smallest capacitance sizes, for which there are many other options. The plate form of capacitor has the considerable advantage that metallized plates can be stacked together to increase the capacitance (Figure 4.4), for very little increase in bulk.

Mica capacitors can be made in single-plate form or with stacked plates. In the past, mica plate capacitors have been made using tinfoil laid between mica plates, or with the plates held together using metal eyelets. These older forms are now obsolete, and the only remaining type is the silvered mica construction which has layers of silver deposited on to the mica, whether the capacitor uses a single plate or multiple plates. The silvered mica type of capacitor has the best combination of electrical, thermal and mechanical properties that can be found for a capacitor of low value.

Natural mica has a relative permittivity value of around 5.4, and this value is maintained up to very high operating frequencies, certainly as far as 1 GHz. The dissipation factor is very low at frequencies of 1 kHz upwards, of the order of 0.0003, although at 50 Hz the dissipation factor is about 0.005 because of the presence of ions

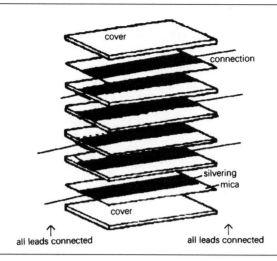

Figure 4.4 Stacking metallized plates together to form a larger capacitance value with alternate plates connected.

in the material (which causes the ruby colour of the natural mineral). The dielectric strength is quite astonishingly high, of the order of 150–180 kV/mm, and this is due to the plate form of the material. The structure of mica is of flat molecules of aluminium–potassium silicate which bond together into sheets which are ultimately one molecule thick. There is no natural conduction path across these sheets, because the distance between sheets is much greater than the distance between molecules along the sheet, so that any conductivity has to be along the sheet rather than from sheet to sheet. Even the thinnest slices of mica that we can cut consist of many sheets, so that the insulation and the dielectric strength are unrivalled by any material in which the molecules are arranged in a three-dimensional structure.

The volume resistivity of natural mica is 5×10^{15} Ωm, which is not the highest of values but which is an average value that does not take account of the enormous differences that are caused by differing directions of measurement. A value of resistivity which is taken in a direction along a sheet of mica will be much less than a value measured between sheets, and the quoted value is an average. Mica is an example of an anisotropic material, one whose physical attributes will vary according to the direction in which a length is taken. All crystalline materials are anisotropic and materials which form flat sheets, like mica, are very noticeably so. This property is not confined to minerals and crystals – wood is an example of a very well-known anisotropic material in which the strength depends on the direction of the grain.

The temperature coefficient of a silvered mica capacitor is positive, and is in the region of $+50 \pm 50$ ppm/$°$C, not so low as typical ceramics. The larger values of capacitance provide the lower values of temperature coefficient. Manufactured silvered micas are available in the range 2.2 pF to 10 000 pF (10 nF), and the usual encapsulation is with wax coated in a ceramic cement. The normal working temperature range is from $-40°$C to $+80°$C (some to $+150°$C or more), with a power factor of 0.002 and insulation resistance of around 10^{10} Ω. Working voltage is usually 350 V maximum, and this rating includes pulse operation.

Silvered micas are now expensive in the UK as compared to other capacitor types (this is not true in the USA), but their combination of parameters is unequalled by any other type, so that applications which call for the highest possible stability must specify these capacitors. Typical applications are tuned circuits and filters for which the stability of frequency is all important. Because of their physical shape, micas have a very low self-inductance, so that their resonant frequency is very high, and the low loss (very low equivalent series resistance) makes the effective Q value (ratio of reactance to resistance) also very high.

All capacitors have a value of self-inductance which is low for the low-capacitance values, but fairly high for some wound foil types. As a result, for each value of capacitor there will be a resonant frequency when the self-inductance is in series-resonance with the capacitance. At this frequency, the capacitor has its minimum impedance, and above this frequency the impedance will be more predominantly inductive. The Q-factor for the capacitor will also be a minimum at the resonance frequency. The physical shape of silvered mica capacitors makes the self-inductance very low, particularly when the capacitors are made in a form suitable for surface mounting (see Chapter 8). Large-value ceramic capacitors and foil types (other than the extended foil types) have comparatively low values of self-resonance.

Ceramic capacitors are, in contrast, very often specified in situations where loss is of little importance. Unlike mica, the ceramics that are used for capacitors are artificially manufactured, although from natural materials. The traditional materials such as magnesium silicate and aluminium oxide have been supplemented by others such as barium titanate and titanium dioxide, and manufacturers tend to use mixtures whose composition and processing are not revealed. Most manufacturers now quote standard specification letters/numbers rather than the precise materials.

Of these standards, the old established N750T96 bears the 750 number because this is its temperature coefficient when formed into a capacitor, and the N signifies that the coefficient is negative. A corresponding N150 material is also available, but the most stable capacitors are manufactured from C0G materials (formerly known as NP0), with zero temperature coefficient and low soakage. All of these types have low loss characteristics and have replaced silvered mica for critical applications.

- Ceramic capacitors of value 120 pF or lower are almost invariable of the C0G (NP0) type.

Many other types of ceramics, particularly those with a high titanium content, have very high permittivity values, as high as 6000 in some examples. Unfortunately, many of these ceramics are also highly anisotropic in a very undesirable way – the value of relative permittivity alters when the applied electric field is altered, so that the capacitance value is voltage variable. Materials such as barium titanate are, in fact, piezoelectric, meaning that the dimensions of the whole crystal will alter as the voltage across the material is altered. A few materials have high relative permittivity that is combined with reasonable stability, and one of the specifications for such capacitors is X7R/2C1. For less demanding applications, where a variation of capacitance value with applied voltage or temperature can be tolerated, the specification Z5U/2F4 can be used.

For some types of ceramic capacitors, the dissipation factor can be substantial, of the order of 0.15% (0.0015) for the C0G/NP0 type, rising to 3% (0.03) for the Z5U type, so that the equivalent series resistance of these types is comparatively high. The C0G/NP0 type, nominally of zero temperature coefficient, can have values of ± 30 ppm/$^\circ$C, which is acceptably low. The other types have much higher temperature coefficients which are variable so that the value of temperature coefficient will itself vary as the temperature is changed. For these capacitors, it is more usual to replace temperature coefficient by a maximum-change percentage. For example, if a ceramic capacitor has the figures +56%, −35% given in place of temperature coefficient, this means that the maximum change that can be expected at the extremes of the temperature range will be of these percentages. The rated temperature range for the X7R material is −55°C to +125°C and for the Z5U is −10°C to +85°C. Typical maximum changes over these temperature ranges are +15% to −25% for X7R and +56% to −20% for Z5U.

The applications for which ceramic capacitors can be used must therefore be tailored to the type of dielectric that is used. Capacitors, mainly in the 10–100 pF range, that use the NP0 dielectric are suitable for general (usually low-voltage) purposes, including oscillator tuned circuits, timing circuits and filters whose specifications do not demand the use of silvered micas. The more stable of the high permittivity materials, X7R, is specified for values up to about 0.1 µF, and these capacitors are used in bypass and decoupling applications, less demanding filtering circuits, timing, and for coupling applications in which the temperature stability is less important. The Z5U dielectric has the highest range of relative permittivity values and is used to obtain very high capacitance values in the range 0.22 µF to 1 µF. These capacitors are used mainly for decoupling and bypass applications, although they can also be used for coupling in circuits whose time constant need not be stable. The insulation resistance of the smaller capacitance value is of the order of $10^{11}\ \Omega$, but for the larger values a formula $10^{9}/C\ \Omega$, with C in microfarads, is used to specify resistance.

- Of all ceramic capacitors, C0G/NP0 types alone are suitable for sample-and-hold circuits. These ceramics are available in sizes up to 0.01 µF.

High relative-permittivity disc ceramics are made specifically for decoupling purposes for both analogue and digital circuitry. Most digital circuits generate very sharp pulses as devices switch on and off, and these pulses can be spread by way of the DC power supply lines or bus lines unless suppressed. In most examples, it is necessary to place a decoupling capacitor at each IC, connected between the positive supply line and ground, but on some circuits using low clock rates this can be relaxed to one capacitor for each five ICs. Stability of value is unimportant in such an application, where the important features are high capacitance in small bulk and low inductance.

Modern disc ceramics are well suited to this purpose, with a capacitance range of 1 nF to 100 nF (0.1 μF). These can be obtained as low-voltage types, suitable for digital circuits, and high-voltage types that are used in TV and radar circuitry. Tolerance of value is large, in the range +80% to −200%, and variation with temperature is seldom quoted. Insulation resistance of 10^{10} Ω is typical. A more specialized shape for digital use is the low-profile DIL type which is of the shape and size of an IC, but flat, with four pins arranged so that two pins will fit into the positive and negative supply positions of typical ICs and the two other pins are dummies. These DIL capacitors can be fitted into an IC mounting position under the IC, so minimizing the inductance of leads, and if necessary can be fitted on top of existing ICs if existing decoupling is inadequate. The pin ranges are for 14-, 16-, 20-, 24-, 28- and 40-pin ICs.

- Note that the older type of disc ceramics had comparatively high self-inductance, making them unsuitable for decoupling in critical applications. The more recent multilayer disc types are much superior.

Ceramic plate capacitors are also used for lead-through (feed-through) capacitors, used for low-pass filtering when a supply cable is taken through a metal panel. Values range from 100 pF to 10 nF, and the combination of series inductance and parallel capacitance can be specified in terms of the decibels of attenuation for high-frequency signals, assuming a standard line impedance of 50 Ω. Lead-through types are not effective for sine-wave signals of less than 10 MHz, but are very useful for digital circuit filtering of supply lines, particularly now that high clock rates of 800 MHz and above are employed in computer circuits. Attenuation values range from 1 dB for 10 MHz/100 pF to 63 dB for 1 GHz/10 nF.

There is also a range of low-permittivity capacitors with negative temperature coefficients, intended for temperature compensation. The principle is that by combining a main capacitor, which has a positive temperature coefficient, in a tuned circuit along with a smaller value which has a negative temperature coefficient, it is possible to cancel the effects of temperature altogether over a reasonable frequency range. As the main capacitor can be a mica type, with a very low positive value of temperature coefficient, only a small value of capacitor with negative temperature coefficient need be connected in parallel; alternatively a large capacitance value can be used connected in series. The dielectrics that are used are the N150 to N750 types, and even the C0G/NP0 type can be used because its temperature coefficient can range from +30 to 30 ppm/°C. Commonly used values range from 2.2 pF to 220 pF, but much larger sizes are available, up to 0.01 µF. Several manufacturers colour code the capacitors in order to indicate what temperature coefficient is applicable.

METALLIZED FILM TYPES

Metallization by evaporation can be applied to materials which are far from being heat resistant, and which would absorb moisture if they were metallized by a chemical method. At one time, paper dielectric capacitors were made using strips of waxed or oiled paper and aluminium foil. These were able to achieve large capacitance values (compared to the silvered mica capacitors that were used at the time) by making use of a large area of dielectric, but the overall size was kept compact by rolling the foils into a cylinder (Figure 4.5). Paper as a dielectric material has now been superseded almost entirely by plastics materials which can be manufactured to much closer specifications, and the aluminium foil has been superseded by aluminium metallized coatings, produced by evaporation. This allows much thinner dielectric films to be used, with the metal in perfect contact with the dielectric, so that higher capacitance values can be achieved in lower bulk, and with better characteristics than ever could be achieved using paper as a dielectric.

• Older capacitors of rolled construction used connecting tabs placed at one end of the foils. This causes the self-inductance to be high, and modern practice is to overlap the foils to either end, so that a metal end-cap makes contact to all of the edge of a foil, Figure 4.5(b).

The plastics that are used are almost as varied as the plastics industry can supply, mainly polyethylene (mainly as polyethylene terephthalate or Mylar) polypropylene, polycarbonate, and

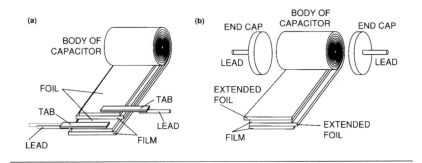

Figure 4.5 (a) The rolled construction used originally for paper capacitors and now for many types of plastic film capacitors; (b) how self-inductance is reduced by making connections to the whole edge of each foil.

polyester, although PTFE (polytetrafluoroethylene), also known as Teflon has also been used, as has cellulose acetate and others. Polystyrene has also been used in the past, but capacitors using this material are no longer in production. None of the plastics materials offers very high values of relative permittivity, particularly polystyrene, so that the reduction in bulk as compared to the traditional paper type of capacitor is due mainly to the use of evaporated metal coatings in place of metal foil.

These evaporated coatings also offer the useful advantage of being self-sealing, in the sense that if a spark-through occurs at a moderate voltage, this will evaporate off some of the thin metal coating and prevent any further conduction. This self-healing effect does not apply at low operating voltages. In place of a fully rolled construction, some plastic film capacitors are available which use interconnected layers of foil or metallizing. These are referred to as multilayered in the UK and as stacked-foil types in the USA; they have the advantage of much lower self-inductance values.

The plastics that are used show considerable variation in performance. The maximum working temperature for many plastics is low, and derating usually has to be applied when temperatures start to approach this limit. In this respect, polystyrene is poor, with a limit of $70°C$ for most types, although a few types permit use up to $850°C$. Polycarbonate types, by contrast, can generally be used up to $100°C$ with little or no derating, although polyester types are generally rated for a working voltage at $85°C$. The temperature coefficients of capacitance are also markedly different, with polyester having a large positive temperature coefficient and polystyrene a smaller negative value. None of the plastics materials has a constant value of temperature coefficient. All of the currently used materials have high insulation resistance values, but the difference between best (polypropylene) and worst (Mylar) can be of the order of 10^3. Insulation resistance values are often quoted in terms of megohm-microfarads, the product of resistance in megohms and capacitance in microfarads. Thus a rating of $10^6 \, M\Omega \, \mu F$ means that a 1 μF capacitor will have insulation resistance of $10^6 \, M\Omega$; a 0.01 μF will have resistance of $10^8 \, M\Omega \, \mu F$ and so on.

The run-of-the-mill Mylar or polyester dielectric capacitor is the tubular axial type encapsulated in epoxy resin, with a typical value range of 0.01 μF to 2.2 μF. These will be used for consumer electronics and for non-critical applications generally, and are available in various voltage ranges, typically 250 V DC (125 V AC) and 400 V DC (200 V AC). These working voltages are quoted for a maximum working temperature of 85°C and although a working temperature range of −40°C to +100°C is usually quoted, the working voltage will have to be derated very considerably for temperatures above 85°C. The power factor is not outstanding, of the order of 0.01 maximum, measured at 1 kHz. The temperature coefficient is positive, 500–600 ppm/°C, and the insulation resistance follows a 10 000 MΩ μF rule.

For more critical applications, particularly in computing and industrial electronics, there are miniature versions of the polypropylene film type which are encased with a shrunk-on sleeve. Working voltages of 250, 500 and 750 V DC (half these amounts for AC) are available, but in a restricted range of values between 0.001 μF and 0.047 μF. These working voltages are quoted at 80°C and as usual, will have to be derated considerably at higher voltages within the working range of −55°C to +100°C. The temperature coefficient is the same at 200 ppm/°C as would be expected, and for this particular format, the insulation resistance is quoted as a minimum of $10^5 \, M\Omega$.

Polyester film capacitors are also available in block form. The principles of construction are either multiple layer (stacked), like the larger values of ceramic or mica, or the use of a long strip (as for the rolled type) which is then repeatedly folded rather than rolled. The casing is then moulded round the capacitor, often using epoxy resin at the sealing lines, although some types use a dip coating of epoxy resin for better sealing. All of these capacitors provide large capacitance values in a small space, and can be obtained in sub-miniature forms with voltage ratings of 50 V DC (30 V AC) where space is at a premium.

The miniature dipped-case type is coated with an epoxy resin which is water-repellent, resistant to solvents and flame retarding as well as being an excellent insulator. Capacitance values of 0.01 µF to 2.2 µF are usual, with 20% tolerance (sometimes 10%). Insulation resistance is better than 10^4 MΩ. Dissipation factor is in the range 0.0075 to 0.015 for frequencies of 1 kHz to 10 kHz. The temperature range is the usual −40°C to +100°C, subject to the usual derating above 85°C. The subminiature versions are intended for low-voltage uses, with an insulation resistance in excess of 2000 MΩ. The commonly used range of values is 0.001 µF to 0.1 µF, all at 20% tolerance, but values outside this range are available. The quoted dissipation factor is a maximum of 0.01. Where pulse use is permitted, the maximum amplitude is usually equal to the DC rated voltage, but the rate of rise or fall is comparatively slow at around 5 V/µs. These capacitors are not suitable for use in circuits where the rate of change of voltage across the capacitor is large.

The encased types of polyester block capacitor cover the value range of 0.01 µF to 2.2 µF, and working voltages of 63 V to 400 V DC, subject to the usual lower AC values, and to derating for high temperature operation. Some types quote temperature coefficients of 300 ppm/°C, others (mainly the lower voltage types) give only the expected range of change as +5%. The larger moulded case types have an insulation resistance of better than 25 000 MΩ; for some of the smaller types this figure can be as low as 15 000 MΩ. Power factor for either range is about 0.01 at 1 kHz.

More specialized types are available for more specialized purposes. The encapsulated polyester block capacitors that are intended for low-voltage telecommunications and industrial applications are made for a rated 30 V, 50 V or 63 V. Since the applications generally call for a working voltage of between 5 V and 20 V, the rating is very conservative, even allowing for any derating for temperature (and rather more derating will be needed for the semiconductors in the circuit). The quoted working voltage is at the usual 85°C with a temperature range of −55°C to +100°C. In the UK, temperature coefficient figures of 200 ppm/°C are quoted, although higher figures of the order of 600 to 800 ppm/°C are always quoted in the USA – this difference is puzzling, particularly when the source is probably the same Japanese manufacturer. Insulation resistance follows the rule of 10 000 MΩ μF.

The multilayer form of construction can be used to create very compact block shapes in a range of high-performance capacitors which have good self-healing properties. Values range from 0.01 μF, 400 V to 2.2 μF, 100 V with 10% tolerance. The temperature coefficient is 200 ppm/°C, with the working voltage quoted at 85°C in a range of −55°C to +100°C. The dissipation factor is 0.01, which is fairly high, but the insulation resistance is better than 10^4 MΩ. These capacitors are often insulated with only a very thin film of plastic, so that they should not be mounted close to other components nor tightly against the PCB.

One other form of construction for the polyester (or other film) type of capacitor with a rolled construction is as a mains-connected type, often in a steel-cased format. This is used where a large capacitance value is needed, and size/weight factors are less important than non-polarized working. In particular, this refers to connection across the mains for power-factor correction or transient suppression, also for reservoir capacitance use in circuits where electrolytic capacitors cannot be specified. The connection to the outer layer of foil is usually indicated by a stripe on the casing.

One form of these capacitors is the steel-encased block type rated at 600 V DC or 250 V AC. These have replaced the oil-impregnated paper types that used to be common in such applications. The casing is not electrically connected, and is present only to ensure containment in the event of an explosive fracture of the capacitor due to catastrophic overload. The capacitor is mounted by using a metal clip or solder lugs, and it should be preferably mounted to a metal chassis rather than to a PCB. The rated voltage is at 70°C and considerable derating may be needed – a 600 V DC rating at 70°C becomes 400 V AC at 85°C, with pro rata changes in the AC rating. Metallized Mylar capacitors should not be operated at low voltages.

The smaller block types are intended for connection across a mains supply in filter circuits, usually for RF suppression, rather than as any part of a smoothing or power-factor circuit. The 250 V AC rating is for a working temperature of 75°C. These are available in two classes. Class X working allows for direct connection across mains leads, but not if failure of the capacitor could cause risk to electric shock to an operator by, for example, making a metal panel live to mains. The class Y capacitors are rated as being suitable for connection across the mains in more dangerous situations in which capacitor failure could cause danger of shock (BS 6201: Part 3 in the UK). The class X types can use polyester film and the class Y types normally use polypropylene film; the metallized polyester type are inherently self-healing at other than low voltages.

The class X capacitors are encapsulated first into epoxy resin and then into flame-retardant casings, and have values of 0.01 µF to 0.47 µF. The working voltage is 75°C, and the derating factor is 1.25% of voltage per °C above 75°C, subject to the absolute temperature range of −55°C to +100°C. The quoted temperature coefficient is 300 to 400 ppm/°C, and the power factor is better than 0.008 at 1 kHz. The insulation resistance is obtained from 10 000 MΩ µF (minimum 30 000 MΩ).

The class Y capacitors come in a value range of 0.0022 µF to 0.047 µF, and can also be used in AC circuits other than mains circuits operated at low frequencies at voltage levels up to 500 V rms. The dissipation factor is better than 3×10^{-3}, and the temperature range is −55°C to +85°C at full ratings. The insulation resistance is shown as 5000 MΩ µF, subject to a minimum of 15 MΩ. A pulse rate or rise or fall rating is also specified; 125 V/µs for the smaller capacitance sizes and 50 V/µs for the larger values.

Polystyrene capacitors are obsolete now that better materials are available. The main drawback of polystyrene was the restricted temperature range of −40°C to +70°C, with the working voltage often being quoted at 25°C. There is also a marked discrepancy between DC and AC working voltages, so that a capacitor which is rated at 160 V DC is often rated as only 40 V for AC. The tolerance can be surprisingly tight (at a price), of the order of 2.5%, and the value range is of the order of 10 pF to 10 nF. Power factors are very low, less than 0.001 at 1 MHz, and the temperature coefficient is in the range −70 to 230 ppm/°C. The insulation resistance is very high, of the order of a million megohms, giving a very long time constant for leakage.

The remaining film-type capacitors are those using polypropylene and polycarbonate. Polypropylene has the lowest power factor ratings of all the plastic films, along with very high dielectric strength which makes it possible to construct capacitors of high voltage ratings. Pulses with high rise or fall rates can be applied, and the material has a consistent value of relative permittivity that does not fall for high-frequency signals. Propylene is used for a full range of capacitors of voltage ratings from 50 V to 1500 V DC, in capacitance values of 0.001 µF to 0.47 µF. These working voltages are quoted at 70°C and the derating factor is 50% for operation at 100°C, the maximum temperature of operation (minimum is −55°C). The temperature coefficient is negative, 200 ppm/°C, and the insulation resistance is very high, better than 10^5 MΩ with a power factor of less than 10^{-3} at 10 kHz. For the 1000 V range, the permitted pulse rate of rise or fall is 1800 V/µs, a rate that can be bettered only by mica and some ceramic types. For the 1500 V capacitor range, the permitted pulse rate of rise/fall is 950 V/µs.

Polycarbonate material is considerably less easy to obtain in thin-film form and so is found only where its superior characteristics demand its use, mainly in military, telecommunications and industrial uses. Polycarbonate types can be in slab form, plastic encapsulated, or for the most arduous conditions, tubular and encased in brass, insulated with epoxy resin. A temperature range of −55°C to +100°C with no derating is specified for the brass-cased types, with a value range of 1 µF to 10 µF. The other varieties can be obtained in a wider range of values down to 0.01 µF. The temperature coefficient can be very low, +50 ppm/°C, and the power factor is better than 0.005 at 1 kHz. Typical working voltage is 63 V DC (45 V AC) for use in low-voltage circuits.

The slab type exists in a larger range of 0.01 µF to 4.7 µF, and in voltage ratings from 160 V DC (100 V AC) to 630 V DC (300 V AC), although use on AC mains is not recommended. The working temperature range is −55°C to +100°C, subject to derating above 85°C, with the same minimum temperature coefficient of ±50 ppm/°C. The insulation resistance is a maximum of 3×10^4 MΩ for the formula 10^4 MΩ µF.

The smaller capacitor types can have their values colour coded rather than printed, although in Europe, it is more common to print value than to use colour coding. A five-band code is used, with the first three bands used for value in picofarads (two significant figures and multiplier). The fourth band is black for 20%, white for 10% tolerance, and the fifth band is red for 250 V DC working and yellow for 400 V DC working. Tantalum electrolytics (see later) are also found colour coded using four bands, but with values in microfarads rather than picofarads and with the fourth band indicating working voltage (but not using the standard number equivalents for the colours).

Electrolytic capacitors

The electrolytic capacitor is a subject by itself, and it has to be treated separately from all other capacitors. The principle is that several metals, notably aluminium and tantalum, can have very thin films of their respective oxides formed on the surface when a voltage is applied in the correct polarity (metal positive) between the metal and a slightly acidic liquid. These very thin films then insulate the metal from the conducting liquid, the electrolyte, forming a capacitor; an electrolytic capacitor. The name comes from the resemblance to an electrolytic (metal plating) cell.

● This same effect causes the problem of polarization of cells, see
   Chapter 7.

The most common type of electrolytic capacitor makes use of aluminium foil, which can be etched, dimpled or corrugated to increase the effective area, enclosed in an aluminium can which is filled with a slightly acid solution of ammonium perborate in jelly form. The capacitor is formed by applying a slowly rising voltage to the capacitor, with the foil positive and the case negative until the voltage reaches its rated level and the DC current falls to a minimum, indicating that the insulation is as good as it is ever likely to be. From then on, when the capacitor is used, it must have a DC (polarizing) voltage applied in the same polarity so as to maintain the insulating film. If the capacitor is used with the voltage reversed, the film will be dissolved, removing any insulation and allowing large currents to pass through the liquid, which will vaporize, destroying the can. The electrolyte is usually in jelly form, but the devastation that can be caused by an exploding electrolytic (not to mention the noise) ensures that no one who has achieved this is willing to try again.

The use of tantalum as the metal of an electrolytic allows for a very different form of construction, in which the oxide film is more

Figure 4.6   Typical aluminium electrolytic sizes (Photo: Nichicon Corp.).

stable and able to withstand reversals of voltage. Tantalum capacitors (*tantalytics*) can be used without a steady polarizing voltage, can be run with the electrolyte virtually dry, and have generally better characteristics than the traditional aluminium type of electrolytic. Experience with the use of tantalum has led to the development of 'dry' electrolytes for the aluminium type of electrolytic also.

- Tantalytic capacitors should not be used in audio coupling applications in which there is little or no bias voltage.

Because of the very fragile nature of the insulating film, which can be only a few atoms thick, electrolytic capacitors are always liable to have a large amount of leakage, so that leakage current at rated voltage is quoted rather than power factor or dissipation factors. Leakage is often related to the capacitance value and working voltage, and the formula:

$$I_{\text{leakage}} = 4 + (0.006 \times C \times V)$$

is often used, with I in µA, C in F and V in volts. For example, using this formula for a 200 µF capacitor at 12 V gives leakage current of $4 + (0.006 \times 200 \times 12) = 18.4\,\text{µA}$. Several manufacturers will make use of this formula to quote leakage values. No manufacturer will *guarantee* an electrolytic to have low leakage value, but the measured values are often surprisingly good if the electrolytic is being operated in reasonable conditions. Bob Pease quotes examples of 500 μF electrolytics with 2 nA leakage at 10 V working.

Many manufacturers also quote a life expectancy in excess of 100 000 hours, at 40°C and rated voltage, for electrolytics, since there is still some prejudice against their use for anything other than consumer electronics. Military applications generally forbid the use of electrolytics, but they are now widely accepted for industrial equipment. Temperature ranges of −40°C to +85°C are often quoted, but considerable derating is needed at the higher temperatures, and there is a risk of freezing the jelly type of electrolyte at the lower temperatures. This is counterbalanced to some extent by an increase in losses as the electrolyte freezes, leading to higher dissipation and subsequent thawing. This, however, is not an effect that you should rely upon. Some types can incorporate vents in order to relieve gas pressure inside the electrolyte.

Electrolytics are used predominantly as reservoir and smoothing capacitors for mains-frequency power supplies, so that their most important parameters, other than capacitance and voltage rating, concern the amount of ripple current that they can pass. For each capacitor the manufacturer will quote a maximum ripple current (typically at 100 or 120 Hz), and also two parameters that concern the ability of the capacitor to pass current, ESR and impedance. The ESR is the effective series resistance in milli-ohms, typically 50 mΩ, for low-frequency currents, and this value may set a limit to the ripple current that can pass; also to the effectiveness of the capacitor for smoothing. The other parameter is effective impedance in mΩ measured at 10 kHz and 20°C, which is used to measure how effectively the capacitor will by-pass currents at higher frequencies. If an electrolytic capacitor is used in a decoupling circuit which is likely to handle a large frequency range, other capacitor types should be used to deal with frequencies higher than 10 kHz, such as a polyester type for the range to 10 MHz and a mica or ceramic for higher frequencies. A useful rule of thumb is to have one electrolytic for five ceramics or discs.

The general-purpose type of electrolytic uses aluminium, often with a separate aluminium casing rated at 1000 V insulation value. The physical form is a cylinder with tag, rod or screw connectors at one end. The capacitance range is generally very large for the lower-voltage units, up to 15 000 μF for 16 V working, but at the higher voltage ratings of 400 V, values of 1 μF to 220 μF are more usual. Many designers avoid the use of an electrolytic at more than 350 V working. The tolerance of value is large (-10% to $+50\%$) and permitted ripple currents range from 1 A to 7 A depending on capacitor size.

- For an exhaustive set of application guidelines for aluminium electrolytics, see the website:

  http://www.nichicon-us.com/tech-info.html

Another useful rule of thumb is that you need 1000 µF of smoothing per ampere of DC output, but this is not necessarily satisfactory. Suppose, for example, that a 5000 µF capacitor is used with a 6 V supply at its full rated ripple current of 5 A and has an ESR of 50 mΩ. The sawtooth ripple will amount to 6 V peak-to-peak, with a further 5×0.05 V $= 0.25$ V due to the ESR, almost negligible. The dissipation in the capacitor will also be too high, and in this type of circuit it is better to use several capacitors in parallel.

Smaller electrolytics are made for direct mounting on the circuit boards for decoupling or additional smoothing, and these are cylindrical and wire terminated, either axial (a wire at each end) or radial (both wires at one end). The voltage range can be from 10 V to 450 V, with a working temperature range of $-40°C$ to $+85°C$ (derating advised at the higher temperatures), and with power factors that can be as low as 0.08 and as high as 0.2. The largest range of values, typically 0.1 µF to 4700 µF, is available for the smaller working voltages. The sub-miniature versions have working voltages that range from 6.3 V to 63 V, and with leakage current which is 3 µA minimum, with the larger capacitance units having leakage given by the formula: $0.01C \times V$. For example, a 47 µF 40 V capacitor could have leakage of: $0.01 \times 47 \times 40 = 18.8$ µA, but measured values are usually much smaller, as low as 10 nA or even less for modern capacitors.

A specialized wet-electrolyte type is made for the purposes of memory back-up in digital circuits. CMOS memory chips can retain data if a voltage, lower than the normal supply voltage, is maintained at one pin of the chip. The current demand at this pin is very low, and can therefore be supplied by a capacitor for considerable periods. This is not the method that is used for calculators, which use a battery, but for such devices as central heating controllers which must retain their settings if the mains supply fails for a

comparatively short period. Typical values for these electrolytics are
1F0 and 3F3. Discharge times range from 1 to 5 hours at 1 mA and
300 to 500 hours at the more typical current requirement of 5 μA,
but the high leakage current must be taken into account.

Solid-electrolyte types are now available in the aluminium range
of electrolytics. Unlike the traditional type of aluminium electrolytic, these need no venting, and cannot suffer from evaporation of
the electrolyte. Also, unlike the traditional electrolytic they can be
run for periods with no polarizing voltage, and can accept reverse
voltage, although at only about 30% of rated forward voltage at
85°C, considerably less at higher temperatures. Typical sizes are
from 2.2 μF to 100 μF, with voltage ratings of 10 V to 35 V at
85°C. Temperature range is −55°C to +125°C, and even at the
maximum working temperature of 125°C, the life expectancy is in
excess of 20 000 hours. The leakage currents are fairly high, in the
range 9 μA to 250 μA, and the ripple current ratings are in the
range 20 mA to 300 mA. One important feature is that the specifications place no restrictions on the amount of charge or discharge
current that flows in a DC circuit, provided that the working
voltage is not exceeded.

TANTALUM ELECTROLYTICS

Tantalum electrolytics invariably use solid electrolytes along with
tantalum metal, and have much lower leakage than the aluminium
types. This makes them eminently suitable for purposes such as
signal coupling, filters, timing circuits and decoupling. The usual
forms of these electrolytics are as epoxy-coated miniature beads or
tubular axial types. The voltage range is 6.3 V to 35 V, with values
of 0.1 μF to 100 μF. The temperature range is −55°C to +85°C.
Tantalum electrolytics can be used without any DC bias, and can
also accept a small reverse voltage, typically less than 1.0 V. A
minimum leakage current of 1 μA is to be expected, and for the
higher values of capacitance and working voltage the leakage
current is found from the capacitance × voltage factor, subject to
the minimum guaranteed value of 1 μA. Power factors in the range
of 0.02 to 0.2 can be expected. Care should be taken not to exceed
the surge voltage rating, typically 1.3 × rated DC voltage rating.

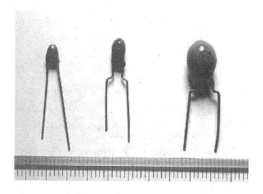

Figure 4.7   Typical tantalum electrolytic packages (Photo Nichicon Corp.).

Variable capacitors

At one time, it would be normal for a catalogue to list variable capacitors of 500 pF maximum or 350 pF maximum in single-, double- or three-gang pattern. Such capacitors, once used in radio receivers, are now available only for replacement purposes and can be quite difficult to find. The reason is the almost universal adoption of electronic tuning methods, making use of voltage-variable capacitors, the semiconductor varicap diodes. The only remnants of the old-style variable capacitors are trimmers.

Trimmers consist of single-turn variable capacitors, using one fixed set and one rotating set of plates, with a solid insulator, usually polypropylene, separating the plates and providing a considerably greater capacitance than would be expected in a small bulk. Capacitance variation depends on the number of plates and size of trimmer, and ranges from 1.4 pF to 10.0 pF and from 5.5 pF to 65 pF. Temperature range is −40°C to +70°C, and the voltage rating is typically 100 V. The dissipation factor is very low, typically 10^{-9}.

Failure mechanisms of capacitors

Capacitors are responsible for a substantial fraction of the failures in service of electronic equipment, with a reliability that is somewhere between that of discrete semiconductors and resistors. It is therefore important to know why capacitors fail and how the reliability of capacitors can be enhanced. This concerns both the designer and the service engineer, because if a unit is known to fail at frequent intervals because of a capacitor problem, the capacitor should be replaced with one that can be expected to have a longer life.

Capacitor failures can be classed as open circuits, short circuits and leakage failures and all three can be caused by the most common of all capacitor problems, internal sparking. Sparking can either evaporate metal contacts, causing an open circuit, or it can penetrate a dielectric, causing shorting. Plastic dielectrics can char under heavy sparking, creating a carbon track which forms a short circuit, although modern plastic dielectrics will permit metal to be evaporated in the usual self-healing mode. Leakage failure is less common, but can be caused by internal short circuit paths or by external paths (around the side of the dielectric).

The main causes of premature capacitor failure are excessive voltage, excessive current or excessive temperature, or combinations of all three. Combinations of parameters can be particularly damaging, because although a capacitor may be rated for operation up to 120°C and for 100 V working, this does not imply that a voltage of 100 V can be used at a temperature of 120°C. Mica and ceramic types are less likely to be affected by running at extremes of their parameters for short periods, but plastic dielectrics have to be treated with much more care, and electrolytics in particular should always be conservatively rated. Sudden charging or discharging can be damaging to capacitors, and is the reason for so many types quoting a maximum rate of rise or fall of voltage.

● The older type of electrolytics used at high voltage levels can fail if they have been out of use for some time and are suddenly subjected to working voltage. This is because the gas film has dissolved, and the recommended procedure is to re-form the electrolytic by gradually increasing the applied voltage to its normal level.

The power factor or dissipation factor for a capacitor is seldom important for DC applications, but when a capacitor is used to handle AC, particularly high-frequency AC, then a poor power factor can cause self-heating because of the current flowing through the equivalent series resistance. Unlike resistors, capacitors are assumed to work at ambient temperature, with no self-heating, and even a small amount of dissipation in a capacitor can be destructive, because the dielectric is also a good heat insulator and will not permit the heat to be easily dissipated. Because of this, the internal temperature can rise well above the external temperature, causing failure. As usual, the plastic dielectric types are particularly susceptible because of the low melting point of most plastics. For electrolytics, the internal resistance at normal working temperature is known, and the dissipation is more easily calculated for the maximum ripple current. Note, however, that when electrolytics are used at unusually low temperatures (but within their permitted range), the internal resistance can rise considerably, causing much greater dissipation, and hence overheating which is often localized and will lead to failure.

One of the prime causes of capacitor failure is locating capacitors near high-dissipation components such as resistors or, in some cases, transformers or chokes. The permitted ambient temperature around capacitors is lower than for resistors, and a good rule is never to subject capacitors to temperatures that would be unsuitable for semiconductors. Assuming that the temperature around capacitors can be controlled, the other factors are then voltage, particularly surges, and current, both of which are reasonably predictable from a knowledge of the circuit in which the capacitor will be used.

Chapter 5

Inductors and inductive components

Induction and inductance

Electromagnetic induction was discovered by Michael Faraday in 1831. The principle is that an EMF (a voltage) is generated in a conductor when the magnetic field across the conductor changes. In the early experiments, the change of magnetic field was accomplished by moving either the wire or a magnet, and this is the principle of the alternator and dynamo. An EMF can also be induced without mechanical movement, when the strength or direction of a magnetic field across a wire is altered, and even the presence of a wire is not necessary, because the alteration in a magnetic field can produce an electric field in the absence of any conductor. Inductive components in electronics make use of the EMF that is generated when a field changes either in the same piece of wire (*self-induction*) or in another piece of wire (*mutual induction*).

The amount of EMF that is generated in a wire can be greatly increased if the wire is wound into a coil, and as much as possible of the magnetic field is guided through the coil. Figure 5.1 shows the flux path in a solenoidal winding for a steady current. Concentration and guidance of the magnetic field is achieved by using a magnetic core, for which the traditional material was annealed

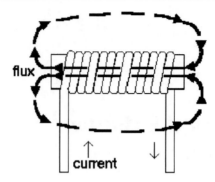

Figure 5.1 The simple solenoid, showing current and magnetic flux paths.

'soft' iron. One way of looking at a soft iron core is as a conductor for magnetism, using the idea of magnetic flux. It is possible to think of magnetic circuits in which magnetic flux (φ) is the counterpart of current, in a path which has reluctance (analogous to resistance), and in which the amount of flux is produced by a magnetomotive force (**MMF**). The equation that is the magnetic equivalent of $V = R \times I$ is then:

$$\mathrm{MMF} = \varphi \times S$$

with φ representing the amount of flux and S the reluctance of the magnetic circuit.

This allows magnetic circuits to be analysed much in the same way as we analyse current circuits, so that if a transformer uses a large iron core with a small air gap then values for reluctance for both the core and the air gap can be calculated, and these will add, as the values of resistors in series are added. The total **MMF** can be calculated from knowing the number of turns of magnetizing winding and the current flowing, so that the amount of flux can be calculated. For any such arrangement, the air gap is the determining factor, acting like a large resistance in a current circuit, which makes the variable values of other resistances in series virtually negligible in effect. The simplest inductive component, other than a straight wire, is a coil of wire with no metal core. A changing current in this coil produces a changing magnetic field, and this in turn induces an EMF in the coil itself. By Lenz's law, this EMF is in a direction that always opposes the change that causes it.

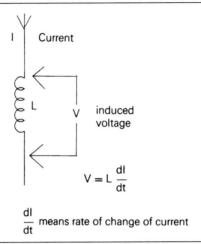

I Current

L

V induced voltage

$$V = L \frac{dI}{dt}$$

$\frac{dI}{dt}$ means rate of change of current

Figure 5.2 Self-inductance, L, defined in terms of the voltage generated by a sudden change of current.

Therefore if the changing magnetic field is caused by an increase of current when a voltage is applied, the induced EMF will be in the opposite direction, reducing the current.

If the changing magnetic field is decreasing because the current has been interrupted, the induced EMF will be in a direction that will increase the voltage at the interruption in the circuit, causing a spark so as to continue the current. Because the EMF is always in a direction that opposes any change of current it is usually referred to as the *back-EMF*. This simple single-coil arrangement has self-inductance, and is a single inductor.

The practical definition of self-inductance is illustrated in Figure 5.2, as the constant L in the equation. When the back-EMF is measured in volts, with rate of change of current in units of amperes per second, then the unit of self-inductance is the henry (H).

Large inductors can be obtained with values of several henries, but for many purposes the smaller units of millihenry (mH) and microhenry (µH) are used. For a coil with only air in its core, or any non-metallic material, and with a length that is large compared to its radius, the value of self-inductance L can be calculated from the dimensions of the coil and its number of turns (Figure 5.3). This value of L is a constant.

$$L = \frac{2.8 \times r^2 \times n^2}{r + (1.11 \times s)}$$

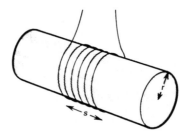

Figure 5.3 Approximate inductance for an air-cored single-layer coil (a sole-noid), with L in µH, s and r in centimetres.

When a core of magnetic material is used, the value of L is enormously increased. The increase is very much greater than the effect of a solid dielectric on the capacitance of parallel plates, and is due to the *relative permeability* of the magnetic material (compare the effect of relative permittivity for a dielectric). The effects are not comparable, however, because whereas relative permittivity of a dielectric is a constant, the relative permeability of a magnetic material is variable – it depends on the value of magnetic field around the materials, and the permeability values that existed previously (the *magnetic history*). The uncertainty of the value of relative permeability makes working with inductors considerably more difficult than working with capacitors or resistors, and for many years the use of inductors in circuits has been declining. The use of ICs has hastened this trend, and devices such as surface-wave filters (see Chapter 8) which make use of acoustic waves have superseded the use of inductors in many types of tuned circuits.

As well as self-induction, we can have mutual induction, in which the rate of change of current in one coil causes an induced back-EMF in another coil. The definition of a coefficient of mutual inductance, M, is shown in Figure 5.5, and since the units of EMF, current and time are the same, the unit of M is also the henry. Mutual inductance is the mechanism of transformers, which are generally bought ready-made, although RF transformers, consisting of two windings on a non-metallic core, are often assembled for one-off applications.

Despite the trend to dispense with inductors there is still a substantial need for inductors of all kinds, particularly for development work. Unlike capacitors and resistors, inductors other than power

Figure 5.4 A typical selection of small inductors using cores.

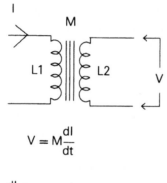

$$V = M\frac{dI}{dt}$$

$\frac{dI}{dt}$ is rate of change of current in L1

Figure 5.5 Definition of coefficient of mutual inductance, M. Mutual inductance is very difficult to estimate and 'cut and-try' methods often have to be used.

transformers are seldom available in the UK as stock items, and most have to be wound as needed on cores whose contribution can be quantified by the manufacturer. Some components catalogues in the UK make little reference to inductors for other than mains or audio uses, but supplies are readily available from the US and other sources. In some ways, inductors are the last lingering

reminder that we have of the early days of radio when the experi-
menter had to make each and every passive component.

Core materials and permeability

The fact that electromagnetic waves can travel through empty space
implies that space has a value of permittivity and also of permeability. The value of the permeability of free space, ε_0, is, in absolute
units, approximately 1.26×10^{-6} henries per metre (compare the
absolute units of permittivity, which are farads per metre), so that
no material will have a permeability value of less than this. The
relative permeability of a material, ε_r, is defined by:

$$\varepsilon = \varepsilon_0 \times \varepsilon_r$$

so that relative permeability is a pure number, a factor that is the
ratio of the permeability of a material to the permeability of free
space. Relative permeability values of as high as 10^6 are possible in
practical conditions. In theory, the relative permeability of any
magnetic material can take any value from almost unity to $+\infty$. In
practice, we can usually find a working average value for a
magnetic core material that can be used to calculate the likely self-
or mutual inductance of a coil or coils respectively.

The difficulty of using permeability values is illustrated by Figure
5.6, which is a typical magnetizing–demagnetizing graph for a
magnetic material. On this type of graph, the magnetic flux density
(field strength) is displayed vertically, and the magnetizing current
is plotted horizontally. If the material is initially unmagnetized,
then the graph starts from the centre and reaches a maximum.
This maximum expresses the saturation field strength – the core
material is fully magnetized and increasing the amount of current
through the coil will make only a very small change to the magnetism. When the current in the coil is reduced from this saturation
value, the graph follows a hysteresis shape, meaning that the path
that is followed for reducing current is not the same as the path for
increasing current. There will be a magnetic flux density when no
current flows through the coil. This is the residual magnetism or
remanence of the core, and for the type of cores that are used for

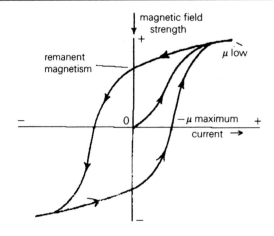

Figure 5.6 A graph of magnetic flux density plotted against magnetizing current. The permeability is represented by the slope of this line, which varies between almost zero and a very large value. Permeability, unlike permittivity, is not a constant.

inductors in electronics, this residual magnetism should be very small, almost negligible.

- One exception in the past (around 1950–1960) was the use of core memory for computers in the dinosaur era, which used small magnetic cores with several windings (usually single turn) as a form of memory, with one direction of magnetization signifying a 0 bit, and the opposite direction a 1 bit.

When the current in the coil is reversed, there will be a point when the reverse current has reduced the new magnetic flux to zero, and when the current is increased further in this negative direction, the flux density will rise to a saturation level in this direction. Returning the current to zero and then to a positive value will then trace the remainder of the curve, but the initial section will not be retraced unless the material is totally demagnetized (usually by heat treatment).

The types of materials that are used as cores for inductors for many types of electronics applications have hysteresis curves that are typified by Figure 5.7, in which the residual magnetism is almost negligible. Such a material is called a 'soft' magnetic material, so called because the first such material to be discovered

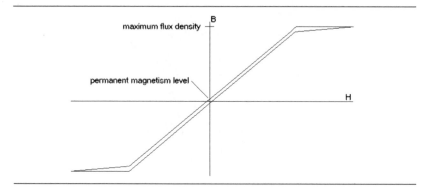

Figure 5.7 Typical idealized hysteresis curve shapes for magnetic materials used for memory cores.

was soft (annealed) iron. Because the shape of the hysteresis curve around the central portion is almost straight, an average value of permeability can be measured for this region and used in calculations. Since this value must always be approximate, and because there is often a quite substantial temperature coefficient of inductance (in the range 100–200 ppm/°C), these self-inductors are often adjustable. Very few suppliers quote the temperature coefficient of inductance for their products. The adjustability is by movement of the core which is threaded and screwed into a plastic former so as to allow the self-inductance value to be adjusted, usually by +20%. The portion of hysteresis curve that is used depends on the amount of any DC that flows in the inductor, so the value of inductance will depend heavily on the amount of such DC, and many inductors are designed for signal use only, or for use with very small amounts of DC only. The effect of large direct currents in an inductor concerns mainly the larger iron-cored inductors, mainly mains transformers, and will be considered later.

Inductive reactance

A self-inductor generates a back-EMF whose size depends on the rate of change of current, so a self-inductor with an applied AC voltage will have an AC back-EMF, which will be out of phase with the applied voltage. The result is that an inductor will pass much less alternating current than its DC resistance would imply, and the current will be in a phase which is 90° lagging behind the applied voltage.

- For some inductors, notably the coil of a moving-coil loudspeaker, the inductive reactance value is completely swamped by resistance at the frequencies that are used, so that inductance is often ignored.

The value of inductive reactance is shown in the table of Figure 1.9 which also shows the formula. Since inductive reactance is proportional to frequency, the values of inductance that are needed in circuits are likely to be considerably smaller for use with high-frequency signals than with low-frequency signals. Each inductor will, however, have a stray capacitance which will cause resonance at some frequency, so that above this resonant frequency, the frequency of maximum impedance, the reactance starts to fall and will become capacitive, with current leading voltage. In addition, the magnetic material itself may be unsuitable for use at high frequencies.

The effect of using an inductor with AC is that there can be an AC voltage induced in the core material itself. This will cause considerable loss, because if the core permits a current to flow because of the induced voltage, then power will be dissipated in the core. Losses of this kind are called eddy-current losses and can be minimized by making the core from thin sheets of magnetic material (a laminated core) or by using a magnetic material which is a very poor conductor (a ferrite core). The latter method is more common for inductors that are to be used for signals, but laminated cores are used extensively for power transformers.

A more serious effect, in the sense that it cannot be so easily avoided, is that of the hysteresis curve. When a core material is repeatedly magnetized and demagnetized, the conditions in the core are changing in a way that is represented by continually going round the hysteresis curve. Each cycle through the hysteresis curve, however, represents a loss of energy, and this appears as dissipation in the core itself, heating the core. The amount of energy lost in this way on each cycle is proportional to the area enclosed by the hysteresis loop (Figure 5.8). The higher the frequency of the signals in the inductor the greater the loss in this way, so that for the frequencies above 50 MHz it is quite common to make use of self-in-

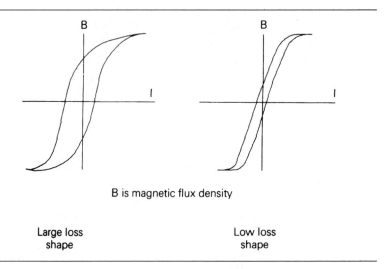

B is magnetic flux density

Large loss
shape

Low loss
shape

Figure 5.8 Hysteresis curves and dissipation. The area enclosed by the hysteresis curve is a measure of the energy dissipated each time the material is magnetized and demagnetized.

ductors with only an air core, although there are a few materials that can still be used at 100 MHz and above.

All of the above omits any mention of the other major cause of dissipation, the resistance of the wire which is used to construct an inductor. The ratio of reactance to resistance for an inductor is never so very high that the resistance can be neglected, so that the power factor for an inductor is never negligible as it so often is for a capacitor. In addition, the DC resistance of a wire may not be the factor that determines dissipation for signals, because at the higher frequencies signal current flows on the surfaces of wires rather than internally, so that the equivalent resistance becomes greater. In the frequency range up to about 100 MHz, inductors are wound with stranded wire, known as 'Litz' wire (an abbreviation of Litzendraht), whose surface area is considerably greater than that of a solid wire of equivalent cross-section. In the frequency range of approximately 100 MHz upwards inductors usually consist of self-supporting coils of thick copper which have been silver plated to increase the conductivity for surface currents. The 'goodness' of any inductor at a particular frequency can be assessed from its Q factor, equal to the ratio of reactance to resistance, but this is an approximation only and does not hold true as the frequency of self-resonance of the inductor with its own stray capacitance approaches.

At the highest frequencies for which separate inductors are used, inductors are more likely to consist of pieces of straight wire or metal strip rather than wound coils, and at such frequencies tuned-line theory is more useful than the resonance of inductor and capacitor. Our uses of inductors as separate components are therefore confined to lower frequencies and to a limited range of applications. In much modern equipment, signals are generated using crystal control and make use of harmonics of the crystal frequency rather than of the variation of a resonant circuit by altering either capacitance or inductance. The main uses of inductors are then as high impedances (chokes) for signals in simple filtering circuits. For more elaborate filtering, it is likely that filters based on resistor–capacitor circuits (for broad bands) or on surface-wave devices (for narrow bands) will be used.

Inductors in the range $1\,\mu H$ to $1\,mH$ can be bought as stock items, with an inductance tolerance of 10%. The inductance values are commonly measured at $100\,kHz$, and typical resistance values range from $0.04\,\Omega$ for the $1\,\mu H$ inductor to $30\,\Omega$ for the $1\,mH$ component. This corresponds to Q values in the range 15–21 at $100\,kHz$, but since reactance increases with frequency so also does the Q value, so that Q values for these inductors are usually quoted at higher frequencies that are still well below the frequencies of resonance. Typical resonant frequencies range from $190\,MHz$ (Q quoted at $15\,MHz$) for a $1\,\mu H$ inductor to $3\,MHz$ (Q quoted at $800\,kHz$) for a $1\,mH$ inductor. The maximum permitted DC is also quoted, but more from the point of view of dissipation than for change of inductance value. The core material that is used is a ferrite type so that eddy current losses are negligible and hysteresis losses are low. Since the values are non-adjustable, these inductors are best suited for use in filter circuits and as RF chokes for power supplies, etc.

Winding small inductors

Despite the availability of ready-made inductors for choke uses, there is still a need to wind inductors to specific purposes. The first decision that has to be made is whether or not to use a core. This can easily be settled by reference to the cores that are available, and very often the only cores that are widely available will be of the general-purpose type intended for frequencies up to about 2 MHz. For appreciably higher frequencies, then, the use of an air-cored coil is more likely, and since formers for coils can be used with or without cores (cores are usually sold separately) the procedure for designing the coil is much the same. The approximate inductance of the coil needs to be known, and if a core is to be used, its relative permeability under the working conditions will also need to be known. If a core is to be used, the inductance of the coil itself will be the required final value (with core) divided by the relative permeability value.

The inductance of a single layer solenoid (a coil whose length is considerably greater than its diameter) is given approximately by the formula in Figure 5.2, of which the most important feature is that inductance is proportional to the square of the number of turns. The importance of this is that if turns have to be removed or added in order to adjust the final (measured) value of inductance then the square law must be taken into account. For example, suppose that a 75-turn coil has been constructed on a former in order to achieve an inductance of 70 µH, but the measured value turns out to be 80 µH. The correct number of turns then has to be found from:

$$\frac{80}{70} = \frac{(75)^2}{(X)^2}$$

Where X is the correct number of turns. In this example, this makes the correct value of X equal to about seventy turns, so that five turns have to be removed. When a coil is wound, it can be useful to keep some spare wire in a straight length in case extra turns have to be added.

The usual problem, however, is to find the number of turns that will have to be wound in order to provide a required value of inductance, and this is always difficult because it requires you to know in advance the radius and length of the finished coil. One method is to make the winding on a standard former, always to the same length, achieving the correct number of turns by using a suitable gauge of wire. This means that the ratio of length to radius can be fixed, and allows the use of the formula illustrated in Figure 5.9.

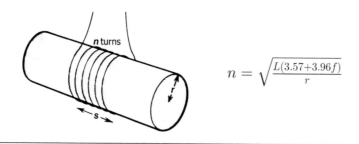

$$n = \sqrt{\frac{L(3.57+3.96f)}{r}}$$

Figure 5.9   Approximate formula for the number of turns required for a given inductance value, assuming a single layer coil. Dimensions s and r are in units of mm and $F = s/r$.

This is an approximation, and for coils whose final value will be provided by the use of a core the approximation should always be close enough to allow the value to be corrected by altering the position of the core. For air-cored coils, the value is likely to be critical only if the coil is part of a tuned circuit, so that correct tuning can be achieved by the use of a capacitive trimmer. For some coils, the required length might be achieved only by spacing the turns, and if uninsulated wire is used (such as the silver-plated wire for high-frequency use) then spacing is essential for insulation. Even spacing can be achieved by winding two lots of wire (which need not necessarily be of the same gauge) together and then un-winding one of the coils.

When a core is to be used, then its relative permeability must be known. This could be a very difficult matter for some materials, but the ferrite materials that are used almost exclusively for signal-current inductors have more reliable values of permeability than the steel-based cores that are used for transformers. Manufacturers of cores deal with the permeability specification in two ways. For the small threaded cores that are used along with plastic coil formers, values of initial and working relative permeability are quoted. The initial relative permeability value is an indicator of the type of material, and is usually the highest possible relative perme-ability value that can be obtained. The working (or coil-permeabil-ity) value is the relative permeability value that can be assumed for use with a coil, and is considerably smaller. Typical values are 250 for initial relative permeability and five for working relative perme-ability. The optimum frequency range for the core will also be

Table 5.1 Typical formulae for enclosed-core inductors – the quantities in the formula will vary from one core type to another and must be obtained from the manufacturer

---

Core factors can be quoted in terms of inductance per turns (squared), or as number of turns for 1mH. Whichever method is used, either a square or a root will be required. Note: 1 000 000 nH = 1 mH.

Example: A core is rated at 250 nH/turn². Find the number of turns needed for an inductance of 2.5 mH.
Divide 25 mH by 250 nH, giving 100 000. Take the square root, which is 316, the required number of turns.
Example: A core is rated at 63.25 turns for 1 mH. Find the number of turns needed for 32 mH.
The required inductance is 32 times 1 mH, and will need $\sqrt{32}$ times the number of turns, which is 357.8. This would be rounded up to 359 or 360.

---

quoted. Use of the coil is not restricted to this range, but it does provide a guide to the region in which the core losses will be lowest.

The alternative approach applies in particular to coil formers which are almost entirely constructed from ferrite material, so that the coil is enclosed by a magnetic path, inside and outside. This makes the effect of the core rather more predictable, and for such coils, the manufacturers of core provide a formula and a set of values which will allow the inductance to be calculated to within quite close limits. The usual type of formula is of the form illustrated in Table 5.1, in which an inductance factor is quoted for each type of core, and the number of turns is found from the formula – note that in such formulae the inductance is usually required in nano-henries, 10^{-9} H. An alternative is to show the number of turns for 1 mH inductance, so that the number of turns for any other value can be found by using the relationship that inductance is proportional to the square of the number of turns. For example, if the number of turns for 1 mH is given as 63, then for 2 mH the number of turns is:

$$63 \times \sqrt{2.5} \doteq 99.6 \text{ turns}$$

This would be rounded to 100 turns. Much higher values of effective relative permeability can be obtained in coils of this enclosed type, typically 100–200. The temperature coefficient of relative permeability will also be quoted; this is usually in the range 150–250 ppm/ °C.

Enclosed coils are particularly suitable for inductors in the 1 mH and above range for frequencies of 1 kHz to 1 MHz approximately. This covers the important audio filter range, along with various modulation and subcarrier frequencies used in telecommunications. The use of adjustable cores allows for tuning with the inductor so that no capacitive trimmers are required. The inductance formulae normally assume that no DC will be applied, and that the signal currents in the coil will not take the magnetic material anywhere near its saturation point. The saturation value of flux density is often quoted, but it is by no means simple to find how this corresponds to current in the coil. One useful guide is:

$$I = \frac{B_{sat} \times A \times n \times \mu_r}{L}$$

Where B_{sat} is the saturation flux density in units of teslas, L is self-inductance in henries, I is current in amperes, A is the area of cross-section of the core in m^2, n is the number of turns in the coil, and μ_r, is the working relative permeability.

Manufacturers often quote the effective area of cross-section for a core in terms of square millimetres, and the values of self-inductance in millihenries, with current in milliamperes, and substituting these units gives the same equation with no correction factors required. For example, for a core of effective cross-sectional area 40 mm^2 and a coil of 1 mH, using sixty-three turns, the saturation flux density is quoted as 250 mT (250×10^{-3} teslas) and effective relative permeability as 170. This makes the saturation current value 630 mA, but this is purely a guide, and it would be unwise to approach anywhere near this value of signal or DC current – a good working guide would be to keep to below 10% of this value. The reason that any more precise calculation is difficult is because there is no exact relationship between inductance value and flux density unless the coil and core are of a very simple geometrical shape (like a ring), and the assumption has been made that the permeability of the core at the saturation flux value is equal to the permeability of free space.

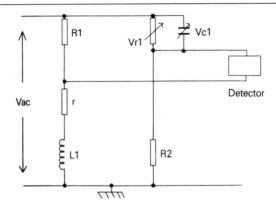

Figure 5.10 A form of measuring bridge for inductance in which an unknown inductance value is compared with a known value.

Measuring inductance

Because winding inductors to precise values is so difficult, the measurement of inductance is important, and the standard method is some form of measuring bridge. This uses a circuit such as that shown in Figure 5.10 to obtain a balance point at which there is no signal at the detector. The value of the inductance and its resistance can then be calculated or, as is more likely in modern equipment, read from dials or digital displays. A bridge measurement is only as good as its standards, so that the bridge should contain standard inductors which are to a close tolerance, although for many purposes, an inductance value that is correct to about 10% is sufficient, since final adjustments will always be made using core movement or by the use of a trimmer capacitor.

The problem here is that many of the low-cost bridge instruments do not allow for measurements on the range of inductors for which measurement is likely to be needed. Small bridges generally have a range of inductance measurement of about 1 mH to 100 H, and measurement is most likely to be needed for the range of 5 µH to 5 mH. The type of automatic bridge, such as the range made by Wayne-Kerr, can cope with measurements in this range with a resolution of 1 nH, and a precision of ±0.5 µH, using a frequency of 1 kHz or 10 kHz. If inductance measurements are the only bridge measurements that are required, and they are not required so often as to justify a costly instrument, then a simple bridge circuit can be used, or a bridge bought which is specifically for the measurement of inductances in the $1\,\mu H$ to $1\,mH$ range. A useful, but slower, alternative is to put the inductor in parallel with a value of capacitance that will swamp stray capacitance values ($500\,pF$ or more) and find the resonance frequency, then calculate the inductance from this value.

Transformers

Transformers make use of the effect of mutual induction, whether they are the multiple winding type of transformer or the autotransformer, in which one single winding is used, with connections tapped for different connections. The main types of transformers that are used in modern electronics circuits are:

1. Mains transformers, used in power supplies.
2. Matching transformers, used for feeding lines.
3. Tuned transformers, used in signal amplifiers to achieve a specified bandwidth.

Of these, the tuned transformers are seldom used now, having been replaced by combinations of wideband ICs and electromechanical filters, so that we shall confine our attentions to the mains and the signal-matching types.

The transformer, other than the autotransformer type, has at least two windings, one of which is designated as the primary winding, the other as the secondary, and the action is that an alternating voltage applied to the primary winding causes an alternating voltage to appear at the secondary. Unless the transformer is intended only for purposes of isolation, the primary and secondary voltage levels are usually different. The conventional style of transformer consists of a bobbin on which both primary and secondary windings are formed, usually with a metal foil layer between the windings to act as an electrostatic screen. The core is then assembled by placing E and I sections of thin steel alloy into place, with the bobbin lying in the arms of the E section (Figure 5.11). There are, however, several other forms of construction. When twin bobbins are used side by side, the electrostatic screening can often be

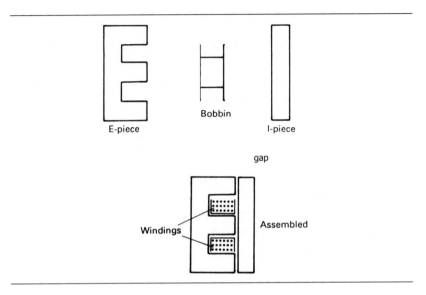

Figure 5.11 The E and I form of core for small transformers.

dispensed with, and some transformers make use of a pair of C-shaped cores rather than the E and I structure. Another form, very common now, is the toroidal transformer, in which both windings are placed over a ring of magnetic material. The toroidal type has in the past been very expensive to produce because of the difficulty of winding the turns into place, but development of toroidal winding machinery has made these transformers much more readily available. Their main advantage is that they have a very low external magnetic field, so that they are often specified for use in equipment where hum pickup must be kept as low as possible.

A perfect transformer can be defined as one in which no power is dissipated, so that the power supplied to the primary winding (primary voltage × primary current) is exactly equal to the power taken from the secondary (secondary voltage × secondary current). Only very large transformers approach this state of perfection, and for the sizes that are encountered in electronics the efficiency of a transformer, defined as:

$$\frac{\text{power output at secondary}}{\text{power input at primary}}$$

will be of the order of 80% to 90%. For many purposes, however, the power loss in a transformer is not particularly important provided it does not cause the transformer to overheat.

Another equivalent definition of perfection in a transformer is that all of the magnetic flux of the primary winding will cut across the secondary winding. This leads to another way of defining transformer losses in terms of 'leakage inductance', meaning the portion of the primary inductance which has no effect on the secondary. Leakage inductance is more commonly used to define losses in a signal-carrying transformer than for a mains type, particularly since irregularities in the response of a transformer to wide-band signals are usually caused by leakage inductance and its resonance with stray capacitances.

As a result of the zero-power-loss definition of a perfect transformer, there is a simple relationship between the voltages and a number of turns at primary and secondary respectively, which is:

$$\frac{\text{primary voltage}}{\text{secondary voltage}} = \frac{\text{primary number of turns}}{\text{secondary number of turns}}$$

assuming that the self-inductance of the primary winding is enough to form a reasonable load in itself, because if the primary self-inductance is too low, the efficiency of the transformer will also be very low. In general, the higher the ratio of reactance to resistance for the primary winding, the more efficient the transformer is likely to be.

For all types of transformers other than autotransformers, the isolation between primary and secondary windings is important. Transformers that are specifically designed for isolation will include a DC voltage isolation test as part of the specification, and for such purposes it is normal for the insulation to be able to withstand several kilovolts DC between the primary and secondary windings without measurable leakage. The insulation from each winding to ground (the core or casing of the transformer usually) should also be of the same order.

Some types of transformers can be used with direct current flowing, and for such transformers the maximum amount of DC is stated, because excessive current could cause saturation. Saturation of the core means that the relative permeability will be reduced almost to the value for air, so that transformer action will be almost lost, and this is usually avoided by having an air gap in the core (see Figure 5.11), thus restricting the amount of flux (see earlier in this chapter). A few transformers are designed (see the telecommunication type, below) so that the core will saturate on overload so as to prevent excessive signal being passed to the secondary circuit, and some types of transformer use the same principle to distort signals for wave-shaping purposes.

Signal-matching transformers

A few types of signal-matching transformers can be bought ready-made. These include 600 Ω line isolating transformers which are used to isolate telephone users' equipment, particularly mains-connected equipment such as facsimile (fax) equipment and computers, from the telephone lines in order to ensure that it would be impossible for mains voltage ever to be connected to the telephone line. Such transformers must be obtained from sources who can guarantee that they are constructed to standards approved by the telephone company; in the UK the relevant specification is HED 25819. These telecommunications isolating transformers are of 1 : 1 turns ratio, and an overload on the consumer's side of the winding will fuse the winding rather than cause high voltages to be passed to the telephone lines. This happens because an overload saturates the core so that it becomes totally inefficient as a magnetic coupling between primary and secondary.

A very common type of signal-matching application is for '100 V line' transformers for public-address systems. Because the power loss of audio signals on long cables is proportional to the square of current, the output of an amplifier for public-address use is usually at a standard level of 100 V for full rated power, so that the current is comparatively low. Since loudspeakers generally have impedances in the 3 Ω–15 Ω range, a matching transformer is needed for each loudspeaker (Figure 5.12). Matching transformers of this type have a selection of secondary tapping points so as to allow the use of loudspeakers of various impedance ratings, and the power handling can be from 1 W to several kW.

● US readers should note that the use of the word 'line' in this context has no connection with power lines.

Another audio application for a matching transformer is the

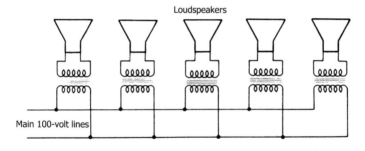

Loudspeakers

Main 100-volt lines

Figure 5.12 Using matching transformers for a PA system.

microphone transformer which is intended to match a low-impedance microphone into a high impedance amplifier. General-purpose matching transformers of this type are designed for moving-coil microphones in the impedance range $20\,\Omega$–$30\,\Omega$, or dynamic microphones in the $200\,\Omega$–$600\,\Omega$ region, and more specialized types can be obtained for ribbon microphones, usually from the manufacturers of the microphones. The primary winding of a microphone is usually centre-tapped so that the microphone cable can be balanced around ground, as illustrated in Figure 5.13, greatly reducing hum pickup, and the whole transformer is encased in metal shielding to minimize hum pickup in the transformer windings.

The other standard forms of signal transformer are pulse transformers, which are intended to transmit pulse waveforms between

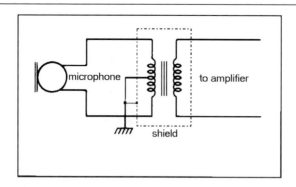

microphone

to amplifier

shield

Figure 5.13 Using a centre-tapped transformer for a microphone lead to minimize hum pickup.

circuits that may be at very different AC or DC levels, such as thyristor circuits. There is no requirement for such transformers to carry low-frequency signals, and their leakage inductance also is of little importance, so very small units can be used, subject to the insulation resistance being sufficient. A typical requirement is for a voltage test to 2.8 kV peak for a transformer intended to work in the bandwidth of 3 kHz to 1 MHz. A factor that is often quoted for these pulse transformers is the voltage–time product, meaning the product of output pulse amplitude (in volts) and pulse duration (in microseconds). This product, typically 200 V µs is a way of ensuring that the transformer does not suffer from excessive dissipation from pulse signals. Pulse transformers of this type can be obtained with $1:1$ windings, $1:1+1$ (two secondaries, or centre-tapped secondary) or $2:1+1$ ratios. Primary inductance levels are in the range 3–12 mH with leakage inductance values of 8–30 µH. These transformers can be obtained as open or fully encapsulated units according to requirements.

For other requirements, particularly RF line to amplifier matching, the transformers have to be constructed to specification. In some cases, a simple tapped winding (autotransformer) will be sufficient; for other applications a transformer may have to be made to a very strict specification. Some of the most useful information on such transformers and on wound components generally is contained in the amateur radio handbooks, either from the RSGB in the UK or the ARRL in the USA. The US manuals have the advantage of containing information on circuits that operate at frequencies and power levels which cannot legally be used by amateurs in the UK.

Mains transformers

Most of the mains transformers that are used for electronics purposes are for power supplies, and as such conform to a fairly standard pattern. These transformers use laminated cores, and the older types use the familiar I and E shaped core pieces which can be fitted together with an air gap. The size of this air gap is a very important feature of the transformer, and is the reason for the difficulties that many users experience when they rebuild a transformer

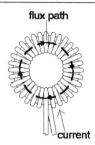

Figure 5.14 The principle of the toroidal winding, which is much more efficient for concentrating flux.

for another purpose, such as rewinding the secondary for a different voltage. The air gap acts for the magnetic circuit of the transformer as a high resistance would in a current circuit, and its magnetic effect is to restrict the magnetic flux in the core. This greatly reduces the likelihood of saturating the core with the large amounts of current that flow in the windings. An air gap is particularly important for mains frequency chokes in smoothing circuits which are likely to carry DC as well as AC ripple, but the use of chokes for this purpose is by now rare.

The traditional I and E, or C, core, however, is not ideally suited to all types of transformer requirements, particularly those which demand a low level of magnetic field around the transformer. A simple solution to the requirement for low external magnetic field is the toroidal transformer, which has become much more generally available thanks to the development of efficient toroid-winding machines in the last twenty years. The main point to note about toroidal transformers is that it can be only too easy to ruin their performance by incorrect mounting, because it is possible to make the mounting form a metal path which is in effect a shorted secondary turn which will dissipate a large part of the energy of the transformer.

The specifications for mains transformers reflect the normal use of such transformers with rectifiers and capacitors to form power supplies. The most important rating is the volt-amp rating (VA) for each secondary winding, expressing the maximum current that can be drawn at the winding voltage. The name volt-amp is used rather than watt because the use of watts would imply a power factor of unity. Because the transformer is not 100% efficient, the volt-amps at the primary will be greater than the sum of the volt-amps at the secondary windings, and part of this, although seldom stated directly, is often implied in a figure for 'magnetizing current', meaning the current which flows in the primary when no load is connected to any secondary winding.

- Modern power supplies make use of active circuits with the aim of keeping the load current in phase with the load voltage and mini-mizing spikes and harmonics. These techniques are beyond the scope of this book.
- A very common practice now is to provide mains transformers with two primary windings rated at 110 V so that the transformer can be used with paralleled inputs on 110 V supplies or with series connections on 220 V.

The regulation of a transformer is an important factor in its use for power supply circuits. When the transformer is loaded by a rectifier and smoothing circuit, and full rated current is being drawn from the secondary (or from each secondary if there are several windings), then the regulation is the fractional drop in voltage, defined as:

$$\frac{\text{open circuit voltage} - \text{full-load voltage}}{\text{open circuit voltage}}$$

and expressed as a percentage. The regulation percentages can be very large for small transformers, typically 20% of a 3 VA type, falling to 5% or less for the larger transformers of 200 VA or more. Some manufacturers quote open-circuit and full-load voltage levels rather than regulation. One important point to watch is that many manufacturers quote the full-load figure for secondary voltage output. This means that for a small transformer with poor regula-tion, the open-circuit voltage can be as much as 20% higher, and al-lowance must be made for this in the circuits which are connected to the transformer. Unless voltage stabilization is used, this order of voltage change between no-load and full-load may be unacceptable for applications that involve the use of ICs.

For any transformer, it is important to have some knowledge of the likely temperature rise during full-load operation. This figure is not always quoted, and an average for the larger transformers is 40°C above ambient for each winding (although most of the temperature rise originates in the secondary windings). Smaller transformers can have greater temperature rise figures, typically 60°C. The maximum acceptable temperature of a transformer is often not quoted and should not exceed 90°C unless the manufacturer specifies another figure. Transformers that use class E insulation can be run at a maximum working temperature of 120°C, but this figure is exceptional among the usual range of transformers for power supplies. The full rating for a transformer implies a 25°C ambient, and the manufacturers should be consulted if higher ambient temperatures are likely.

Since transformers are subject to high peak voltages, the sum of AC and DC voltages, there is a figure of proof voltage (otherwise known as flash test voltage) for each transformer type which is at least 2 kV. This measures voltage breakdown between windings and also between each winding and the metal core. The higher grades of transformers will be tested to higher proof voltages, typically at 5 kV sustained for one minute, and transformers that are intended for special purposes such as heater supplies to cathode ray tubes whose cathodes are operated at very high voltage (negative voltages) will have to be tested to considerably higher voltages. The low voltage requirements of modern instrument CRTs, however, imply that such transformers are seldom required now other than for servicing of old instruments.

The winding resistance of a transformer is not often quoted, although secondary winding resistance is an important factor when designing a power supply whose regulation (before the use of a stabilizing circuit) needs to be known. Note that transformers intended for 60 Hz supplies should not be used in 50 Hz applications. Where winding resistance values are quoted, both primary and secondary will be quoted, and a typical primary resistance for a 240 VA transformer is 4 Ω, with higher values for the smaller transformers. Secondary resistances for low-voltage windings are much lower, of the order of 0.05 Ω for a winding rated at 10 A, higher for windings of lower current rating or for high-voltage windings.

The overwhelming majority of transformers for power supply use have secondary windings that are rated for voltages up to 20 V rms, although because secondaries are often wound with a view to bi-phase rectification (see below), a 20 V secondary will actually consist of two 20 V connected windings, i.e. a 40 V winding with a centre-tap. The output voltage and current that can be obtained

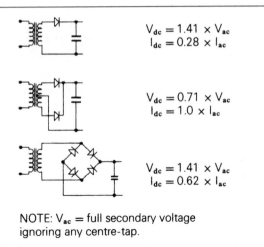

$$V_{dc} = 1.41 \times V_{ac}$$
$$I_{dc} = 0.28 \times I_{ac}$$

$$V_{dc} = 0.71 \times V_{ac}$$
$$I_{dc} = 1.0 \times I_{ac}$$

$$V_{dc} = 1.41 \times V_{ac}$$
$$I_{dc} = 0.62 \times I_{ac}$$

NOTE: V_{ac} = full secondary voltage
ignoring any centre-tap.

Figure 5.15   The standard rectifier filter circuits and approximate performance formulae, ignoring rectifier losses, ripple, and resistance losses.

depend on the form of rectification and smoothing circuits that are used. When the rectifier circuit is followed by a large reservoir capacitor (capacitive input filter), then the DC output voltage of the circuit is high but the current regulation is poor. When a choke input filter is used (the rectifier circuit is followed by a series inductor) then voltage output is lower but current regulation is better.

Figure 5.15 shows the usual standard rectifier and filter circuits, along with the relationship between AC output voltage and DC output voltage and between AC current and DC current. Only capacitor input smoothing has been shown here, because the use of an inductor immediately following rectification (inductive input filter) is very unusual nowadays. The inductive input filter has advantages of better regulation, but the size, cost and weight of the inductor makes the system less attractive, particularly when a voltage stabilizer is likely to be used in any case. Most power supply units use the bridge rectifier arrangement along with a capacitive input filter (reservoir capacitor). When a capacitor input filter is used, the capacitor must be rated to take the full amount of ripple current. As a rule of thumb, the ripple current can be taken as the difference between the AC current and the DC current.

For unusual secondary voltage requirements, it is possible to buy transformer kits, in which the primary winding is supplied on its bobbin, but the secondary has to be wound, and then the bobbins assembled on to the core. These transformer kits are usually of the conventional E and I core type, but several manufacturers supply toroidal cores with a primary winding already provided, and these are particularly useful for very low-voltage supplies which require only a few turns of secondary winding. For each size of core, the manufacturer will quote the number of secondary turns per volt of output, typically from two turns per volt for the 200 VA size to six turns per volt for the 20 VA size.

The wire provided in these kits is the conventional enamelled copper, and the range of diameters is around 0.2 mm to 2.0 mm. When you select a wire gauge for a secondary winding you should bear in mind the power dissipation heating that you can expect at full rated current. For applications needing more than 10 A you will need to use wire of more than 2.0 mm diameter.

- Remember the rule of thumb that you need at least 1000 μF of reservoir capacitor per ampere of output current. Remember also that the rms current in the transformer windings is substantially more than the output DC current.

For details of transformer kits in the UK, see the ElectroComponents catalogue, or the international website at

http://www.rs-components.com

UK users can go directly to

http//rswww.com

You can register at the website to receive information about components, and updates of the product list and technical information.

Other transformer types

Isolation transformers use a 1 : 1 winding ratio and are intended to permit isolation from the mains supply. One important application is in the servicing of the older type of TV receivers, in which one mains lead was connected to the metal chassis. Although this ought to be the neutral lead, there can be no certainty of this, particularly when a twin lead is used with no colour coding. By using an isolating transformer, the whole chassis can, if required, be grounded, or it can be left floating so that there is no current path through the body of anyone touching any part of the circuit unless another part of the circuit is touched at the same time. Isolation transformers are also used for operation of power tools in hazardous situations (outdoors and in very humid surroundings), and some types can be bought already fitted with the standard form of splashproof socket for outdoor use, along with a ground-leakage contact breaker.

Autotransformers consist of a single tapped winding, so that they offer no isolation, unlike the double-wound form of transformer. Fixed ratio autotransformers are intended to allow the use of electrical equipment on different mains voltages, for example the use of US 110 V equipment on European 220 V supplies. The demand for this type of transformer in the UK tends to be localized around US air bases, but there is a large amount of test equipment in use which demands a 115 V supply and which has to be supplied by way of an autotransformer. It is important to ensure that any such equipment cannot under any circumstances be accidentally plugged into 220 V mains, and a fuse should be incorporated to prevent damage in the case of an accidental overload. Autotransformers can also be used to provide 220 V for European equipment being used in a country where the supply voltage is 110 V AC.

The more common type of autotransformer is the variable type, such as the well-established Variac (trademark of Claude Lyons Ltd). This consists of a single toroidal winding with the mains supply connected to one end and to a suitable tap (the taps provide for different mains voltage levels), and an output terminal which is connected to a carbon brush whose position on the winding can be varied by rotating a calibrated knob. This allows for an output to be obtained whose voltage can be smoothly varied from zero to a voltage greater than the mains supply voltage, typically 270 V. Current ratings range from 0.5 A to 8 A depending on the size of toroidal core that is used. Variable autotransformers can be obtained either in skeleton form, with virtually no protection from the windings or connections, or in various degrees of enclosure. Since these are autotransformers there is no mains isolation, and if isolation is needed, it must be provided by a separate isolating transformer used to feed the autotransformer.

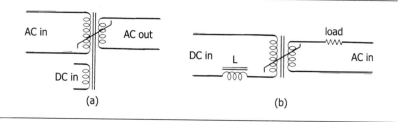

Figure 5.16 (a) Principle of saturable reactors. (b) Control of AC in a load by the direct current through a saturable reactor.

Magnetic amplifiers

The characteristics of a material being magnetized, shown in the drawing of Figure 5.6, can be used as the basis of an amplifier. Normally, for transformer action, we try to use a core material where the characteristic resembles as closely as possible a straight line, with minimum change of slope and smallest possible area of loop. It is possible, however, to make core materials which permit a wide range of characteristics, including some which reach saturation (the point at which the magnetization due to the core has reached maximum). An inductor wound on such a core is called a 'saturable reactor', and its inductance will vary according to the extent to which the core is saturated, Figure 5.16.

If a transformer is wound on such a core, together with a third winding, then, with no current flowing through the third winding, normal transformer action will take place provided that the core is not saturated by the primary current. If DC is now passed in the third winding, the core will come closer to saturation, the primary inductance will be lower and the coupling less so that the output will drop. With the core completely saturated, the output is almost zero. This can be made the basis of an amplifier circuit which consists of reactors only. The operating power is AC, and the 'signal' controlling the amplifier is DC or very low frequency AC. Very high gain figures can be achieved, for a low-voltage passing current in the control winding can control a high-voltage high-current AC, which if necessary can be rectified and used as the DC control signal in another amplifier. Magnetic amplifiers are extremely reliable and are used in industrial control circuits where speed of response is not important, particularly in motor control

circuits. They are generally bought as a complete package, sometimes designed to order by the transformer manufacturer, and seem
to be almost unknown outside the specialized fields in which they
have been used.

Switch-mode supplies

The traditional approach to the design and construction of a stabilized supply is in many ways far from ideal. To start with, a large
amount of energy is wasted because the unstabilized output level
has to be maintained considerably greater than the stabilized level,
and the voltage difference will result in considerable dissipation of
heat from the stabilizer. At low voltage levels in particular, very
large values of capacitance are needed for the reservoir, and this
demands electrolytics of $100\,000\,\mu F$ to $500\,000\,\mu F$, with all of the
problems that attach to electrolytics. The low frequency of the
ripple from a mains-operated supply is always difficult to remove
completely, even using large capacitors, unless the voltage level is
high enough to allow inductors with their inevitable series resistance
to be used.

The answer to many of the problems of making high-current low-
dissipation low-voltage supplies is the use of switch-mode supplies,
which exist in several types. Although these circuits contain both
active and passive components, their action makes use of several
types of passive components in a way that needs to be understood if
you are involved in specifying or servicing such equipment. The
types that are used for applications such as TV receivers use the
mains at full voltage to provide power for an oscillator whose
output in turn is rectified and stabilized. Another option, used particularly for digital equipment, is to employ a step-down transformer to provide mains isolation, rectifying the output of this
transformer to use as the supply to the switch-mode circuits. When
this latter approach is used, the circuit can provide for a step-down
or a step-up of the DC voltage applied to the chips, using an
inductor when a step-up is needed. These circuits also generally
make use of higher frequencies for operation.

The more modern form of switch-mode power supply is illustrated
in the block diagram of Figure 5.17. The mains voltage is rectified,
using a bridge set of diodes connected directly across the mains,

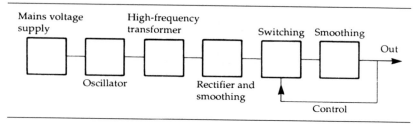

Figure 5.17 A block diagram for a modern form of switch-mode supply, using a high-frequency oscillator and transformer.

and this output is partially smoothed. The rough DC is used to operate an inverter oscillator whose output is connected to a high-frequency transformer – a typical frequency is 50 kHz so that the transformer can be a comparatively small and light component. All stages up to and including the primary of this transformer are live to mains.

At the secondary of this transformer, low-voltage windings can be used, and these will not be live to mains. The output from a winding can be rectified, using a bridge circuit, and smoothed using comparatively low-value capacitors. For low-current lines, a conventional IC stabilizer circuit can then be used, but more usually a switching circuit will be used, chopping the DC into square waves whose mark–space ratio (ratio of high-voltage time to low-voltage time) can be controlled by a voltage-controlled oscillator. This square waveform is again smoothed into DC, and the level of this DC is used to supply the control voltage for the switching circuit. In this way, stabilization is carried out with minimal loss of power, because the switching circuits will be either fully conductive or fully cut off, and changes in the stabilization conditions do not cause large changes in dissipation.

Heat-sink requirements are negligible, and the design of the circuit makes it easy to include cut-off provisions in the event of overvoltage, overcurrent or overheating. In this form, however, the circuit contains some redundancy in the sense that smoothing is being done twice and the square wave formation is being carried out twice, once in the inverter stage and again in the switching stage. The benefits are precise control of low-voltage high-current supplies, with smaller and lighter components (particularly inductors and capacitors), and with ripple at a frequency which is easily smoothed.

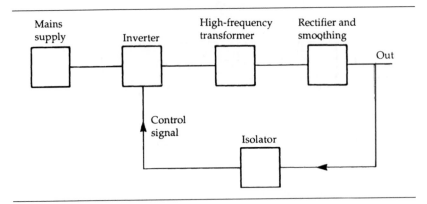

Figure 5.18 Another form of switch-mode supply in which control is exerted on the oscillator by way of feedback through an opto-electronic link.

An alternative, still preserving the mains-voltage approach, is to control the inverter part of the circuit which operates at mains voltage. Figure 5.18 shows the block diagram for a circuit which uses an opto-isolator circuit for this purpose. The usual mains-level bridge rectifier and reservoir circuit operates an inverter whose mark–space ratio can be controlled by a DC signal. The output of the inverter is converted to the correct voltage level using a high-frequency transformer as before, and this is rectified and smoothed. The smoothed output is used to control the inverter by way of an opto-isolator acting to supply the DC control voltage from the inverter using as its input the DC output from the stabilized voltage. In this way, the output side of the circuit is not at mains voltage, but can still be used to exert control on the inverter circuit which is at mains voltage.

This approach is comparatively simple in block form, but because of the losses in the opto-isolator it can require rather more circuitry than might appear to be needed. Another problem is that some regulatory bodies concerned with electrical safety do not consider opto-isolators as suitable for total insulation between mains supply and the chassis of electronic equipment. The problem is that most opto-isolators have only a small separation between input and output, less distance than is stipulated to be used between live parts in electrical safety regulations.

The use of pulse-transformer methods is an alternative that can more easily pass electrical safety approval tests, since a transformer can be manufactured to any required standard of isolation. In addition, the use of a pulse-transformer method obviates the need to have an inverter working at mains voltage, and substitutes in its place a pulse amplifier.

One disadvantage of all switch-mode circuits, however, is RF interference. The essence of a switch-mode supply is that large charging and discharging currents are likely to flow, and where these currents flow in stray capacitances and inductances there is likely to be resonance that can result in RF being generated. The problem is tackled firstly in the layout of the circuit by ensuring that grounding, particularly single-point grounding, is correctly carried out, and circuits are encased in metal screens to reduce radiation as far as is practicable, but from then on, filtering is required to reduce the level of conducted RF interference (RFI) to a minimum. Filtering is made considerably easier if the circuits use a mains-frequency transformer, since this prevents the spread of RFI through the mains lines. The use of a mains transformer with a grounded screen between primary and secondary, along with a 10 nF capacitor across the output terminals of the transformer, results in a reduction of conducted RFI to well below the limits imposed by regulations of bodies such as the FCC in the USA. Further reductions can be achieved by using an inductive filter on the secondary side of the mains transformer, and by filtering each output of the power supply. The usual range for testing for RFI is 10 kHz to 30 MHz, but good filtering will ensure that low levels of RFI are found for frequencies well above the 30 MHz limit.

Chapter 6

Inductive devices

Relays

Electromagnetic relays make use of the magnetic field in the core of an inductor to alter the setting of a mechanical switch. There is an overlap in the meanings of the words relay and contactor; in general a relay is a comparatively low-voltage low-current device, and a contactor is designed to deal with much higher levels of voltage, current or both.

- Contactors are not generally considered to be electronic components and therefore do not feature in this book. For details of contactor construction and use you should consult a specialized text of electrical engineering.

The essential parts of either device are a coil with a soft magnetic core, a moving armature, and a set of contacts which are actuated by the movement of the armature, and which are insulated from the coil and from the metal frame of the relay. The coil circuit is known as the *primary circuit* of the relay, and the contact circuit as the secondary. The principles of the traditional form of relay are illustrated in Figure 6.1. Relays can, with some redesign of the magnetic circuit, be used on AC, making use of the principle that an

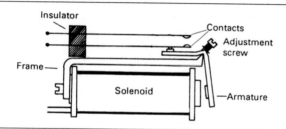

Figure 6.1 Working principles of the traditional type of relay. When the solenoid is magnetized it attracts the armature, and the lever action then closes the contacts.

armature will move so as to close a magnetic circuit when the energizing coil is driven by AC.

- You might argue that since a relay can carry out a power-amplifying action it is not strictly speaking a passive component. From the point of view of this book I take passive components to mean electronically passive, not using thermionic amplification or the action of semiconductors.
- Many of the applications for the traditional type of relay have been filled more recently by the use of reed relays (see later, this chapter) and by the use of semiconductor devices such as thyristors, which are active devices and therefore outside the scope of this book. Devices using optical coupling and triac switching are generally listed under the heading of *relays* in catalogues.

In the mechanical type of relay, the power that is dissipated in the operating coil will determine the mechanical force that is available. This is sometimes specified as *pull-in power*, the minimum power that is needed to make the contacts operate. The *holding power* is also sometimes specified, the power that is needed to keep the contacts switched over after they have been pulled in. There is usually a significant difference between the two values, so that if a relay is intended to remain activated for a long period after initial activation, then the dissipated power can be considerably reduced by lowering the applied voltage or current so that only the hold-in power is dissipated. A maximum power dissipation figure will also be specified which must not be exceeded if overheating is to be avoided, and in high ambient temperatures this maximum value will have to be derated. The range of power levels is sometimes quoted in terms of operating voltage for a particular coil, so that a relay nominally intended for 12 V operation might be assigned a voltage range of 9–13 V. Where relays are intended for AC operation, nominal volt-amp ratings are shown as well as the usual voltage and current ratings for the coil.

- Many uses for mechanical relays depend heavily on the isolation that can be achieved between the primary and the secondary circuits. Some wiring regulations prohibit the use of solid-state relay equivalents on the grounds that such isolation is more difficult and that fail-safe operation (the secondary circuit remains open when a fault occurs) is less certain. These objections can now be overcome, but while the prohibitions remain in force they must be respected.

The figures for test voltage and insulation resistance are therefore important. The test voltage relates to a flash-test, a high voltage applied for a specified time, usually one minute, in which time no signs of breakdown should be visible when the relay is operated at normal temperature, pressure and humidity. A typical test voltage is 2 kV. The insulation resistance is measured at a high voltage, less than that used for the flash test, and typically 500 V. Insulation resistance of 100 MΩ minimum is normal, and some applications call for 10^{12} Ω insulation performance. This level of insulation resistance can be achieved, but *not* if the construction makes use of nylon (whose insulation resistance decreases dramatically in damp and warm conditions).

For a mechanical component, the life, as measured by the expected number of operations before failure, is important. Often only the mechanical life is quoted, typically 10^7 operations. When both electrical life and mechanical life are quoted, the electrical life is usually less by a factor of 10–100. This reflects the main limitation on the life of a relay, which is damage to, or degradation of, the contacts rather than total failure of mechanical operation. The operating temperature range is restricted compared to other electronic components, usually −20°C to +60°C. A few relay types specify maximum mechanical shock in terms of acceleration in 'g' units ($g = 9.81$ ms^{-2}) while the relay is operating, and this part of the specification may be of importance in industrial applications.

The coil resistance of a relay is usually available as part of the

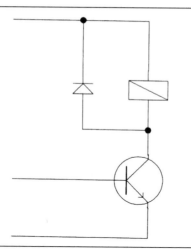

Figure 6.2 Protecting semiconductor components when the coil of a relay is driven by the collector terminal of a transistor or other semiconductor.

printed specification, but figures for coil inductance are harder to come by. The importance of coil inductance arises when the DC current through the coil is controlled by any form of switch, whether mechanical or electronic, because when the coil current is switched off there will be a reverse-EMF whose value will be given by:

$$E = L\frac{\mathrm{d}i}{\mathrm{d}t}$$

Where L is inductance and $\mathrm{d}i/\mathrm{d}t$ is the rate of change of current through the coil, which will be rapid when the coil is being switched off. The inductance of most relay windings is substantial, and a relay coil should never be driven from a semiconductor collector circuit unless there is diode protection, using the circuit shown in Figure 6.2. Emitter connection may be permissible, subject to the ratings of the semiconductor device. Where the relay coil is controlled by a mechanical switch, diode suppression of back-EMF is not strictly necessary, and a series resistor–capacitor suppression circuit can be used. Manufacturers of relays will advise on the suppression circuits that are most appropriate for any particular relay.

Relay contacts deserve particular attention. Contact configurations are specified in terms of the number of poles and whether these are normally closed (NC), normally open (NO) or change-over (CO). For some applications of relays, the contacts spend most of their time closed with current passing; in other applications the contacts are open for most of their lives, exposed to what may be a corrosive atmosphere. For some applications, contacts must have as low a resistance as possible when closed; for other applications the resistance is much less important than lack of corrosion and damage when the contacts make and break at frequent intervals. An ideal contact material would have very low resistance, would need very little force exerted between the contacts to keep the resistance low, would be robust, resistant to corrosion, durable and cheap. There is no single ideal material, and so the contact materials that are used must be selected for the particular application.

In general, contact ratings for DC will be much lower than for AC. This is because the most destructive effect on contacts is *arcing*, a spark which is drawn out into a spluttering discharge or flare as the contacts open, and which will cause metal to be transferred from one contact to another or oxidized. Arcing under AC is of short duration, because the alternating current drops to zero twice per cycle and because the reversal of voltage means that material will not always be moved in the same direction. The effect of arcing is so serious that most relays are rated to use no more than 24 V DC at their maximum current as compared to 240 V AC at the same current. These AC ratings, however, assume that the load will be resistive, and considerable derating will be needed for a reactive load.

The amount of the contact resistance depends on the area of contact, the contact material, the amount of force that presses the contacts together, and also in the way that this force has been applied. For example, if the contacts are scraped against each other (a *wiping* action) as they are forced together, then the contact resistance can often be much lower than can be achieved when the same force is used simply to push the contacts straight together.

In general, large contact areas are used only for high-current operation and the contact areas for low-current relays as used for electronic circuits will be small. The actual area of electrical connection will not be the same as the physical area of the contacts, because it is generally not possible to construct contacts that are precisely flat or with surfaces that are perfectly parallel when the contacts come together. This problem will be familiar to anyone who has renewed the ignition points in a car in the days before electronic ignition circuits. Since the size of the relay determines the amount of contact pressure that can be used, and the area of contact is rather indeterminate, then the main factor that affects contact resistance is the material of the contacts themselves.

The relay designer has the choice of making the whole of a contact from one material, or of using electroplating to deposit a more suitable contact material. By using electroplating, the bulk of the contact can be made from any material that is mechanically suitable, and the plated coating will provide the material whose resistivity and chemical action is more appropriate. In addition, plating makes it possible to use materials such as gold and platinum which would make the contacts impossibly expensive if used as the bulk material. It is normal, then, to find that contacts for switches are constructed from steel or from nickel alloys, with a coating of material that will supply the necessary electrical and chemical properties for the contact area.

The choice is never easy, because the materials whose contact resistance is lowest are in general those which are most readily attacked chemically by the atmosphere. In addition, some materials which can provide an acceptably low contact resistance exhibit 'sticking', so that the contacts do not part readily and may weld shut. The other main problems are burning and oxidation. The spark current which passes at the time when contacts are opening can cause melting of the contacts or cause the metal to combine chemically with the oxygen in the atmosphere (oxidation or 'burning'). Of these two, oxidation is a more severe problem than melting, because the oxides of most metals are non-conductors, so that oxidation causes a large rise in insulation resistance, even to the point of making the contacts useless. This is the usual form of electrical failure for a relay.

The usual choice of materials is illustrated in Table 6.1. From the point of view of resistance alone, silver is the preferred material, since silver has the lowest resistivity of all metals. Unfortunately, silver is very badly corroded by the atmosphere, particularly if any traces of sulphur dioxide exist (where coal or oil is burned, for example), and silver contacts will have a short life if there is any arcing at the contacts.

Silver contacts are desirable if the lowest possible contact resistance is needed for high-current use, but the action of the switch

Table 6.1 Materials for relay contacts. Each has its own particular merits and drawbacks

Material	Advantages	Disadvantages	Uses
Silver	low resistance	corrodes readily	high current and high contact pressure
Palladium-silver	not readily contaminated	high resistance	general use
Silver-nickel	resistive to burning, sticking	high resistance	general use
Tungsten	hard, very high melting point	easily oxidized	high-power contacts
Platinum	stable, unaffected by chemicals	low current only	specialized relays
Gold	can be plated resists chemicals	low current only	electronics equipment

will have to be such that the contact pressure is high, and the contacts are wiped as they are brought together. Silver plated contacts can be used with fewer problems when the relay is sealed in such a way as to exclude atmospheric contamination, or if the contacts are surrounded by inert gases, but these are not simple solutions to the corrosion problem. Some relays are now available which use the compound silver cadmium oxide for contacts. This, being an oxide already, resists further oxidation, but has reasonably low resistivity and is fairly soft·so that good contact can be made at low levels of holding force.

Gold plating is a very common solution to the contact problem, particularly for low-current electronics use. The contact resistance can be moderately low, and the gold film is soft, so that moderate pressure can result in a comparatively large area of contact. Of all metals, gold is about the most resistant to corrosion, although the combination of hydrochloric acid and electric current can cause gold to be attacked quite rapidly, making this material unsuitable if the atmosphere contains traces of chlorine or hydrochloric acid. Since passing electric current through sea-water causes chlorine to be generated, gold plated contacts can be severely corroded when the switches are used at sea. The contacts are suitable only for the lower currents, but since most relay applications in electronics are for low currents this is no handicap. Gold plating of contacts is often mandatory in the specification of switches for military contracts.

Contact coatings based on the 'noble' metals are often employed. These metals are so named because of their high resistance, like gold, to chemical attack, and typical metals in this group are platinum, palladium, iridium and rhodium. All of them have rather high contact resistance, but are very stable and resist chemical attack. Platinum is particularly useful for contacts that will be used for low currents and for high voltage levels. Rhodium and iridium platings provide a high level of resistance to corrosion along with stable contact resistance, and are suitable for medium voltage, medium current levels. Some advantages can be gained by using thick films of contact materials which are alloys of the noble metals with silver. For example, palladium–silver has a much better resistance to contamination than silver itself, although with a higher contact resistance than silver. It is one of the general contact materials that can be used in relays for various types of applications.

The metals tungsten and molybdenum, although not of the platinum group, are also used as contact materials for special purposes. Tungsten in particular is very resistant to melting caused by contact arcing, and is used for high power switching. Its disadvantage is that the surface will oxidize easily, causing contact-resistance problems. The most common general-purpose contact alloy material, however, is nickel–silver. The cost of nickel–silver is such that it can be used as a bulk material rather than as a coating, and although its contact resistance is higher than that of pure silver, it is much more resistant to chemical attack than the pure metal. It also resists burning and the contacts do not tend to stick together.

Where contacts can be hermetically sealed, mercury wetted contacts can be used. As the name suggests, the contact materials are amalgamated with mercury (meaning that the metals dissolve in the mercury) which softens the materials and makes it very easy to obtain a large contact area at low pressure, greatly reducing the contact resistance. Mercury wetted contacts also eliminate contact bounce (see later) and are extensively used in reed relays. Mercury wetting must never be used for open contacts because the mercury will evaporate, and the vapour is hazardous.

The make and break times for relays are often quoted. These are the times in milliseconds that are needed for a change in energization (coil switched on or off) to cause the contacts to change over. The make and break times (or operate and release times) are sometimes identical, but more often the make time is longer than the release time because the time for the collapse of the magnetic field is faster than the time to establish the field, and also because the

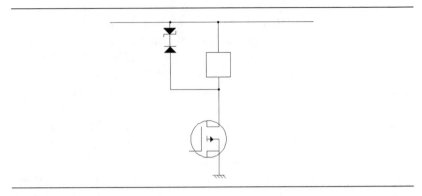

Figure 6.3   Improving relay release times by using a zener diode to catch back-EMF.

make time includes the time for which the contacts bounce. Times of 5–20 ms are typical for the range of small relays used in electronics circuits.

- The release times can be improved where a relay is driven by a semiconductor circuit if a zener diode is used to catch the back-EMF, as illustrated in Figure 6.3.

Contact bounce is a problem that applies to any type of mechanical contacts. No simple mechanical contacts deliver a single voltage change as they close because the elastic nature of the contacts means that contacts will make, then bounce apart and remake at least once, delivering a multiple pulse. Contact bounce in analogue circuits will cause noise, mainly impulse noise, but in digital circuits it can cause total failure of the action, since if the closing of the contacts generates a signal, then bouncing will generate multiple signals. Digital circuits rely on counting the number of pulses rather than the voltage level of these pulses, so that the signals created by bouncing contacts amount to spurious signals in the system. Since the bounce is almost unavoidable (although mercury wetted reed switches, as noted above, are almost bounce-free) the methods that are used to eliminate the bounce must be electronic, using some form of debouncing circuit based on either a flip-flop or the use of a Schmitt trigger circuit.

An older use for relays is as a chopping device. In this application, the relay coil is switched at a mains frequency rate, typically 50 Hz or 60 Hz, and the contacts are connected to a steady voltage, so that the output is a square wave. This type of circuit was at one time used for such diverse purposes as supplying AC for car radios (when 100 V+ supplies were needed), or for DC amplification. The car radio applications called for a mechanically tuned self-oscillating type of relay, and switching noise ('hash') was always a problem. These types of applications became obsolete when semiconductor circuits replaced them.

Relay varieties

Conventional mechanical relays exist in a huge variety of forms, from sub-miniature types that are enclosed in metal cans of the same size as the TO-5 transistor casing, to heavy-duty types that require to be mounted on a steel chassis. Of this range, a few examples will be described in detail as an indication of what can be expected. The smallest relays in a TO-5 can are double-pole change-over types, which can be obtained with coils rated for 6 V, 12 V or 24 V, the standard relay operating voltages, along with a nominal currents of about 15 mA to 60 mA; the lower voltages require higher currents so that the power is constant at about 0.36 W. Contact ratings are typically 28 V, 1 A DC, with a mechanical life in excess of 10^7 operations. The operate and release times are each 4 ms, and the contact resistance is 100 mΩ. The capacitance between open contacts is quoted at 0.7 pF, making this type of relay very well suited for switching high-frequency signals. If you need to switch RF, you should look for a relay intended for this purpose, whose voltage standing-wave ratio (VSWR) is quoted.

Many types of relay are intended for direct mounting on PCBs, and a good example is the miniature double-pole change-over type. The usual range of voltages (6, 12, 24) is provided in three coil choices, with a working power dissipation of 0.7 W. The contacts are silver and rated at 30 V DC or 250 V AC at 5 A. Note the very large difference between the AC and the DC ratings, reflecting the effect of DC on arcing between contacts. The proof voltage is 2.5 kV 50 Hz AC, and the maximum ambient temperature is +55°C. The mechanical life is quoted as being in excess of 10^7 operations, with an electrical life in excess of 2×10^6 operations at full rated load. The whole relay is protected with a clear cover.

Plug-in relays exist in a huge variety of types and sizes, but one very durable series has made use of the type of sockets that were once used for radio receiving valves. These relays use either 8-pin or 11-pin bases, and are also covered, using a clear polycarbonate material which can be removed if needed. This range of relays is available in the usual DC range of operating voltages, but also for 115 V DC, and in AC form for the voltages 12 V, 24 V, 48 V, 115 V and 230 V AC. The power rating for the DC types is 1.3 W, and for the AC types 3 VA. The contacts are rated at 30 V DC or 250 V AC at 10 A, with the contact material being gold-coated silver. Operate time is 15 ms maximum, and release time is 10 ms maximum. The test voltage is 2 kV at 50 Hz, and specified insulation resistance is 100 MΩ. Electrical life is quoted as being in excess of 106 operations, with mechanical life in excess of 10^7 operations. The ambient temperature limit is 65°C when the relay is working at nominal voltage, but only 35°C when the relay is used at its maximum permissible voltage of about 8% higher than nominal voltage. The ambient temperature can be as high as 105°C when the relay is kept unenergized.

A variation of these plug-in relays is a type with a test button and indicator system. Pushing the test button allows the contacts to be changed over manually, with a flag indicator showing that the contacts have, in fact, changed. In addition, a neon or LED indicator (depending on the relay voltage range) will indicate when the relay is energized. This type of relay can be employed usefully when relays are used only for emergency purposes and it is necessary to check the action at intervals and to know when the relay has been triggered.

Heavy-duty relays can be arbitrarily defined as relays which have contact ratings exceeding 10 A, and which are intended for bolting to a steel chassis. A typical heavy-duty 20 A relay has double-pole change-over (DPCO) contacts using silver and silver cadmium oxide surfaces. The 20 A rating applies to 20 V DC or to 240 V AC, and for AC contact operation, the electrical life is quoted as exceeding 104 operations. Coil operation can be DC, using 12 V or 24 V, with a power rating of 2.5 V. For AC operation, voltages are 110 V or 240 V, with a 3 VA rating (hold) and 5 VA to cause change-over (the *inrush* rating). This is by no means the largest current rating that can be obtained, but represents the highest current handling that would normally be considered in terms of electronics as distinct from electrical applications. Relay-like

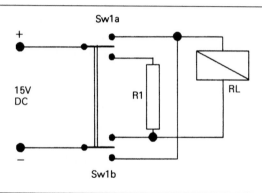

Figure 6.4 Remanence or latching relay circuit, using one button to close the relay and the other to open it.

devices that handle appreciably larger currents and voltages are classed as contactors.

Having looked at these examples which are typical of the range of relays, we can also look at some less common types which are sometimes specified for equipment. The remanence or latching relay is a type that depends on the use of a core material that retains its magnetism to some extent. When current is passed through the coil, the contacts change over, and will remain in that condition when the current is broken. To change the contacts back again, a reduced amount of current has to be passed through the coil in the reverse direction. The manufacturer will specify a suitable value of resistor to connect in series with the coil when current is to be passed in the reverse direction. The remanence relay is a compact and reliable way of implementing a system in which one button is pressed to switch on and another to switch off, and Figure 6.4 shows a typical circuit. The maximum time for which current can be passed is important, and will be specified for a range of temperatures. Minimum time for applying current in either direction is of the order of 10 ms.

The latching relay achieves the same type of action, but by using a permanent magnet and two separate coils. A voltage applied to a coil for at least 50 ms will cause the contacts to change over, and the contacts will remain in this set position until the other coil is activated for at least 50 ms. A latching relay is inevitably larger and more costly than its normal counterpart, but the same specifications as regards contact ratings, proof voltage, insulation resistance and life apply. The coils are usually rated for continuous operation although in practice they are energized only for brief intervals.

Relays can also be obtained with specialized contact systems. An example is the coaxial relay in which the contacts form part of a 50 Ω coaxial system, allowing a change-over action for signals at frequencies up to 450 MHz. The normal relay specifications apply here, and in addition the voltage standing-wave ratio (VSWR) is specified as better than 1.1 : 1, with cross-talk between the outputs at −39 dB.

For electronics purposes, many relays would be used in circuits where the coil operation was by current from a semiconductor device such as a transistor or IC. To obviate the need to build relay-driving circuits into the equipment, advantage can be taken of relays that incorporate driving circuits, transistorized relays. These are available as normal and high-sensitivity types, with the high-sensitivity types using a Darlington type of connection. The usual 12 V and 24 V operating voltages are used, but because of the use of transistors, the range of operating voltages can be very much larger, so that a nominal 12 V unit can operate with a voltage range of 5–25 V, though the advised variation is 11–13 V. The operating currents are in the range 5–10 mA for the standard type, and below 1 mA for the high-sensitivity type. The protecting diode is already wired across the coil, and since the input is to a transistor base circuit, no protection is needed for the driving circuit.

Although like transistorized relays they are not strictly speaking passive devices, solid-state relays can also be bought (as distinct from discrete thyristors). These are available in a set of standard packs, ranging from DIL, PCB and plug-in to surface mounting. A particular advantage is that no mechanical contacts are involved, so that DC ratings are quoted for currents up to 40 A, although they are not suited for very low DC voltages. The devices are particularly suited to switching AC loads, although for inductive loads a combination of a 100 Ω resistor and 0.1 μF capacitor is required with some types to suppress transients. A varistor wired across the output is also recommended for transient suppression in other applications. Some types need to be bolted to a heat-sink for use at full rated voltage and current.

Table 6.2 Reed relay contact arrangements

Form	Contacts	Abbreviation
A	single pole	SP
C	single pole change-over	SPCO
Double A	double pole	DP
Double C	double pole change-over	DPCO

Reed switches

Reed switches are widely used for signal switching, particularly at the lower signal frequencies. The main attraction of the reed type of switch is that the contacts are enclosed and therefore cannot be damaged in adverse environments. In addition, the switch is magnetically operated either by a solenoid coil or by a permanent magnet, so that remote control is possible and variable capacitance to the hand of an operator is not a problem. In this respect, the reed switch is normally used as a secondary switch or relay in the sense that the operating current to its coil will normally be controlled by another switch. Reed switches are generally classed as relays for the purposes of cataloguing electronic components.

The reed switch can therefore be mounted directly on a PCB deep inside the circuit, possibly in an inaccessible place, and operated from any conventional type of low-voltage switch on a panel. This avoids the problems of bringing a signal cable out to a panel-mounted switch and back again. Table 6.2 shows the international designations for reed relay contact arrangements in terms of Form A or Form C.

The basic reed principle is illustrated in Figure 6.5. The thin metal reeds are sealed into a glass tube, and bent so as to make contact because of the effect of a magnetic field. The usual arrangements of contacts are normally open or change-over, and the reed portion can often be specified separate from the actuating coil or magnet. The use of the solenoid coil allows relay type of operation, but the use of a permanent magnet allows mechanical operation (by moving the magnet to or from the reed tube) which can be custom-made to whatever pattern is needed. The typical magnet distances are of the order of 7–11 mm to operate the reed, 13–16 mm to release, so that a movement of around 5–20 mm

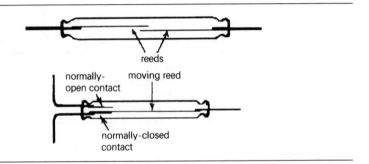

Figure 6.5 Reed relay principles. The thin nickel-alloy reeds are encased in glass and bent into contact by a strong external magnetic field.

would normally be used to ensure reliable opening and closing of the reed circuit. The predominant use of reeds, however, is with the solenoid, using up to 100 ampere-turns to operate the reed.

One important feature of reed-switch use is the fast switching time. This will be of the order of 1–2 ms to make, and can be as low as 0.2 ms to break, with reed resonant frequencies in the order of 800–2500 Hz. Fast switching speed is not necessarily an important feature of signal switches, but if signals have to be sampled or if there has to be switching from one signal source to another at frequent intervals then the reed type of switch has considerable applications. An important further consideration for high-frequency or fast rise-time signals is the very low capacitance between open contacts, which can be less than 1 pF for the normally-open type of reed. The use of silver- or gold-plated contacts allows low contact resistance values of around 100 mΩ, and for applications that demand very low and stable contact resistance, mercury wetted reeds can be obtained. The mercury wetted type also has the advantage of providing very low-bounce operation, a point that will be taken up later. The normal dry contact type of reed is as prone to contact bounce as any other mechanical switching device.

The other features of reed switches are high insulation resistance (because of the use of glass encapsulation), of the order of 100 000 MΩ, and high breakdown voltage of at least 200 V.

- Beware of packaged units of reed switch and coil, because these are usually cased in nylon, which has a greatly reduced insulation resistance in humid conditions and at high temperatures.

The operating currents are also quite surprisingly high for a device with small contacts and limited contact force, of around 0.25–2 A. This is mainly because the reed type of relay does not permit a good magnetic circuit to be constructed; most reed operating coils make no use of any magnetic core material. Mechanical life, depending on type of use and reed construction, ranges between one million and one hundred million operations. Like any other form of contact device, the contacts need to be protected by a diode if they are to be used to switch current from an inductive load, and any device that switches the reed operating coil will also require this protection.

Solenoids

Solenoids are used to convert electrical energy into mechanical force, so that mechanical equipment can be electrically controlled. Solenoids can be obtained for mechanical action only, or ready coupled to devices like valves for liquid or gas control. The specifications for the general-purpose type of solenoids will include both electrical and mechanical ratings, and the solenoids can be classed according to the type of mechanical action as push–pull, lever or rotation.

Another concept that is not encountered for other components is duty cycle. Some solenoids may be required to operate continuously for long periods, so that their duty cycle is 100%. Others may operate only at intervals and for a short on time, so that a solenoid that was activated for 30 seconds in each 5 minutes would have a 10% duty cycle. The ratings of a solenoid include allowances for different duty cycles, so that higher voltages can be applied, if required, to solenoids whose action is subject to a low duty cycle.

Push–pull solenoids are available as DC (12 V or 24 V) or AC (110 V or 240 V) operated devices, with a specified coil dissipation such as 10 W or 11 VA for a 100% duty cycle. The amount of force that can be exerted then depends on the stroke (the amount of mechanical movement) and the duty cycle, assuming that the maximum operating voltage will be used for each value of duty cycle. These relationships are not simple, and are best expressed in graphical form as Figure 6.6 illustrates. The mechanical equipment

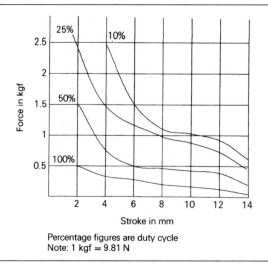

Figure 6.6 Typical force–current graphs for small solenoids. For lb units, use 1 kgf = 2.2 lbf.

that is operated by the solenoid should be arranged so that the combination of force and stroke that the solenoid can deliver will be suitable. The fact that the maximum of force can be delivered at the start of an action (when the stroke is negligibly small) means that the solenoid is well equipped to overcome initial friction. One point to note is that if AC operation is used, the mechanism must not be capable of stalling; the armature must always be able to travel fully into the solenoid, otherwise an excessive current will be drawn.

Lever solenoids are based on relay construction, and are often little more than a relay from which the contacts have been removed, leaving the armature and the frame. These types of solenoids are intended for a range of lower forces and strokes, as illustrated in Figure 6.7. Like a relay, lever solenoids are equipped with a return spring, although this can be removed if there is a spring action in the mechanism that the solenoid is operating.

The rotary type of solenoid produces torque rather than force, but has the same form of graph of torque plotted against rotation for each value of duty cycle. Spring return is normal, but the spring can be removed if there is any other way of rotating the shaft back when the coil is de-energized. The maximum stroke is of the order of 45°.

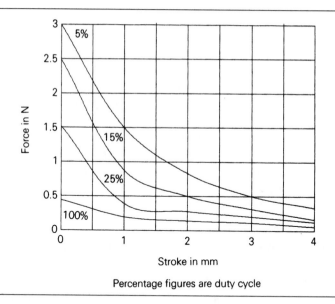

Stroke in mm

Percentage figures are duty cycle

Figure 6.7 Lever-arm solenoids and typical characteristics. For lbf units, take 1 N = 0.22 lbf.

Motors and selsyns

Even a fraction of the vast range of electric motors cannot necessarily be regarded as being electronic components, and in this respect, only these motors which will be actuated by electronic circuits can be considered. Of these, the miniature general-purpose DC motors, using permanent magnet fields, can be included on the grounds that with voltage requirements of 6–12 V and currents of 20–400 mA they can be controlled by the output from a comparatively low-dissipation semiconductor amplifier. Small electric motors of this type can be used in models, rotating signs, display systems and to some extent in robotics.

A much more important type of motor, however, is the servo type of motor which is intended to run in conjunction with a tachogenerator or a potentiometer as part of a closed loop system in which the potentiometer or tachogenerator provides one input to a comparator amplifier (Figure 6.8). A servomotor for this type of application should have a rotor with a very low moment of inertia so that

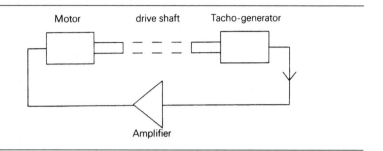

Figure 6.8 A simple servo system. The motor, mechanical connection, generator and amplifier constitute a negative-feedback loop.

rapid starting and stopping can be achieved. This points to the use of an ironless rotor, with the additional advantage that with no high-permeability materials in the rotor there is no tendency for the rotor to stop in preferred positions (called 'cogging') or to have difficult starting positions. In addition, the motor runs more smoothly when it is free of the alternate strong and weak forces that occur as the poles of an iron rotor revolve. The use of precious metals in the commutator and the brushgear also makes for operation at much lower voltages than the conventional carbon-brush type.

Any DC electric motor functions as a DC generator as its rotor revolves, and a motor will reach a stable speed when the back-EMF plus the voltage drop across the resistance of the rotor coil equals the applied voltage. For a motor with a permanent-magnet field or a parallel (shunt) field winding, this means that the speed of the motor is to some extent self-regulating, because when the motor is loaded, the speed will drop, lowering the back-EMF and allowing more current to flow through the rotor coil in order to provide more power. For small servo motors, the back-EMF can be calculated by using figures provided by the manufacturer. One such figure is the EMF constant, which is defined as the back-EMF per 1000 rpm rotational speed. This is usually simpler to work with than the EMF per radian per second factor that is used in textbooks. The EMF constant is usually equal to the rated voltage divided by the no-load speed in kilo-revs per minute.

The parameters that are measured for a servomotor are illustrated in Table 6.3, along with a set for a typical 12 V motor. Most are concerned with the torque and the speed of the motor,

Table 6.3 The quoted parameters for a servomotor and set of typical values

Motor parameters

input voltage	no-load speed	stall torque
power output	no-load current	back-EMF constant
rotor inductance	regulation	terminal resistance
torque constant	moment of inertia	mechanical time constant

Thermal constants (rotor and stator)

thermal time constant	thermal resistance
guaranteed starting voltage	temperature range
max. armature temperature	

Typical set of characteristics for 12 V motor

Characteristic	Symbol	Value	Units
input voltage (max.)	$V_{in}(max.)$	12	volts
no-load speed	ω_0	$7400 \pm 8\%$	rpm
stall torque (start torque)	M_d	88	$Nm \times 10^{-4}$
power output (max.)	$P_{out}(max.)$	1.7	watts
no-load current	I_0	8	mA
back-EMF constant	K_e	$1.6 \pm 8\%$	$V/10^3$ rpm
rotor inductance	L	0.9	mH
regulation		88.9	$10^3/Nms$
terminal resistance	$R_{m(22)}$	$20.8 \pm 8\%$	Ω
torque constant	K_t	$153 \pm 8\%$	$Nm \times 10^{-4}/A$
moment of inertia	J_m	2.1	$kgm^2 \times 10^{-7}$
mechanical time constant	τ_m	19	ms
thermal time constant, rotor		7	s
thermal time constant, stator		550	s
thermal resistance, rotor	θ_{rb}	9	°C/W
thermal resistance, stator	θ_{ba}	26	°C/W
guaranteed starting voltage		0.15	volts
temperature range		−30 to +65°C	°C
max. armature temperature		100	°C

but there are two thermal constants, the thermal time constant and the thermal resistances. The thermal time constant is the time in seconds for the rotor or stator to reach 63% of its final temperature for a given amount of dissipated power. Like resistor–capacitor time constant, this implies that the motor parts will reach their final temperature in about four time constants, so that for a 7 second thermal time constant, final temperature will be reached

Table 6.4 The main motor equations and the steps in determining the choice of motor size and calculating operating conditions

Voltage, current and speed –	V – applied voltage
$V = IR + K\psi$	I – rotor current
Power input –	R – terminal resistance
$P = M\psi$	K – torque constant
Electrical power dissipated –	ψ – rotational speed
$P = I^2 R$	M – torque

Steps in designing servomotor applications:

1 Determine torque required for load. If this exceeds 50% of stalled torque for motor, a reduction gearbox will be needed.
2 Knowing load speed required, calculate power. Choose motor to suit this level of power.
3 Calculate on-load current.
3 Calculate watts dissipated in motor.
5 Calculate temperature rise from thermal resistance figure.

in 28 seconds. The stator time constant for a conventionally constructed motor that uses a massive stator and a light rotor will be very much longer, of the order of 9 minutes. The thermal resistances are expressed in the usual units of °C/W, so that the rise in temperature per watt of dissipation can be calculated. These assume no form of heat-sink.

Temperature is particularly important for servomotor operation, because quantities such as the current for zero-load depend very heavily on temperature. Stall torque and armature resistance have a temperature coefficient value of about 0.4%/°C (positive for resistance, negative for stall torque). The most important calculation is the power requirement, taken from measurements on the load that the motor is to drive. If the power required is not within the capability of the motor a larger motor will have to be used, or the mechanical part of the system will have to be redesigned. Table 6.4 shows some of the basic motor equations and the steps that are required in calculating the current that a small motor will take. For details, the manufacturer of a suitable motor should be consulted.

Whereas the servomotor is a DC device, the selsyn (or synchro) is an AC device which depends on phase differences. The principle is illustrated in Figure 6.9. The rotor is a single-phase winding revolving inside a three-phase stator, and the construction for both

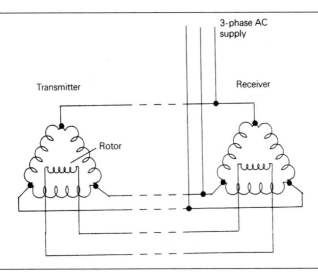

3-phase AC
supply

Transmitter

Receiver

Rotor

Figure 6.9 Principle of the selsyn system for transmitting angular position information.

transmitter and receiver is identical. In use, both rotors are energized from the same AC source (often from an inverter so that the supply is at 1 kHz), and the stator phase leads are connected together. Once energized, the two rotors will take up positions that correspond one with the other, and revolving one rotor (the transmitter) will result in the other rotor (of the receiver) also revolving at the same rate. The two rotors will remain perfectly in step as long as the torque loading on the receiver is negligible, but if the receiver has to drive any significant load, there will be an angular error whose amount depends on the size of the machines and the amount of loading.

Selsyns do not provide the same amount of precision as a servomotor system, nor can they provide the same driving torque, but their simplicity, with no brushgear, no amplification and no solid-state components, makes them very attractive to some light-load applications. One notable use of the principle was to radar displays in which the revolving aerial drove the rotor of one selsyn and the stator coils at the receiving end were used as the deflection coils of a CRT whose display was of a line. As the aerial rotated, the rotating field at the deflection coils caused the line scan of the CRT also to rotate, generating a circular scan (the plan-position indicator (PPI) scan).

Another electromechanical device in the motor family is the stepping motor. As the name suggests, the shaft of a stepper motor will rotate by a fixed angle for each input, and combinations of stepping motor and lead-screw drive can be used to provide linear steps in place of steps of rotation.

The principle of stepper motors is multiphase drive, and a typical stepper would normally use a four-phase set of windings. The stepping action is achieved by pulsing the four windings in sequence, using pulses derived from an IC driver. For example, if the four windings are L1, L2, L3 and L4, then a typical sequence would be:

L1	L2	L3	L4
On	Off	On	Off
Off	On	On	Off
Off	On	Off	On
On	Off	Off	On
On	Off	On	Off

Note that two windings change in each step. The cycle as shown here repeats after four steps, and each step would typically take the shaft through $1.8°$, so that fifty repetitions of this pattern would turn the shaft through a full revolution.

Stepping motors are usually operated at low voltages, 3 V or 5 V DC, and with fairly low currents in the range 100 mA to 2 A. The driver circuits will need to use heat-sinks, however, and very great care must be taken to ensure that none of the leads to the motor becomes disconnected while the other leads are carrying pulses, as this will always cause serious damage. Although the average currents are fairly low, the peak currents in each pulse can be large, so that wire of at least 24/0.2 size needs to be used between the driver board and the motor. If large torque values are required with smaller steps, gearboxes can be attached, but precision would be lost if this is done because of the inevitable backlash in the gearbox.

Magnetic clutches

The magnetic clutch is a useful electromagnetic component which is rather more specialized than most of the components which have been described in this book, and is not usually listed among electronic components. The principle is that a mechanical drive can be made or interrupted by the clutch according to the state of an electrical supply and for the miniature types of magnetic clutch the two main methods of action are powder or liquid. In either case, the action is very similar – two magnetic disks face each other and a magnetic powder or liquid is placed between them.

One disk is connected to a shaft which is powered, and the other disk is connected to an independent shaft, the driven shaft. The whole assembly is surrounded by an energizing coil. When the coil is unenergized, the powder or liquid transmits very little torque to the driven shaft. How much torque is transmitted in this state depends on the packing of the powder or the viscosity of the liquid. When the coil is energized, however, the powder or liquid starts to behave as a solid, locking the disks together so that the driven shaft is to all intents and purposes connected directly to the driving shaft. The torque that can be transmitted is considerably increased if the disks have small blades rather than being completely smooth.

Magnetic clutches offer a fast and smooth take-up of a mechanical drive without the need to start and stop a motor, with all the attending problems of current surge. Characteristics of magnetic clutches vary widely, some are designed to be on/off devices, with no real 'slipping' state, others are intended to allow an amount of slip that varies according to the energizing of the coil. One attraction of the system is that a magnetic clutch system driven by an AC motor offers a variable-speed drive system that does not depend on DC motors nor on the use of thyristors.

AC motors can now be obtained with variable-speed driving circuits. These use either variable frequency, variable pulse width, or a combination of both methods, and are widely available.

Filters and suppressors

A filter is a device which will conduct (pass) one band of frequencies and reject others, the frequencies of the stop-band or attenuation band. An ideal filter would not attenuate signals in the pass-band and would cause infinite attenuation of all signals in the stop-band, with a sharp division between the bands at a frequency called the cut-off frequency. On this basis, the idealized filter characteristics for the main filter types would be as shown in Figure 6.10.

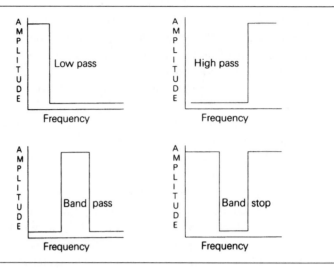

Figure 6.10   Idealized characteristics for the main filter types.

Purely passive filters can be constructed from combinations of inductors and capacitors (LC filters), and from combinations of resistors and capacitors or resistors and inductors. Filters (of very precise characteristics) can also be made using active semiconductor devices such as op-amps, switched capacitors and SAW devices; all beyond the scope of this book.

The use of resistors in a filter, however, means that there will be considerable attenuation in the pass-band, and no sharp separation between pass-band and stop-band, so that RC and RL filters are useful mainly when they are used in conjunction with semiconductors as active filters. The closest approach to ideal filtering is achieved by transducers of the surface-wave type, noted in Chapter 8. Filters can be low-pass, high-pass, band-pass or band-stop, as indicated in Figure 6.10.

The use of inductors and capacitors in filters gives rise to much more complicated response curves than the ideal shapes of Figure 6.10. In addition, the calculation of filter response and of the sizes of components is far from simple. Fortunately, computer software is available for filter calculations, and can be downloaded from several sources on the Internet – use a search engine of your choice to find a phrase such as *filter calculations*.

The two main types of response for a ladder network filter are called Butterworth and Tchebycheff (also spelled as Tchebychev or

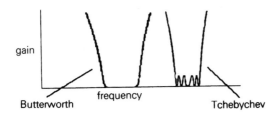

Figure 6.11 The form of Butterworth and Tchebychev responses.

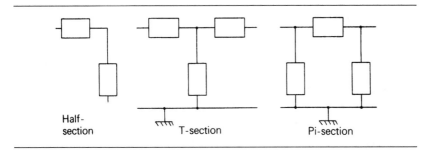

Figure 6.12 The T and Pi arrangements for filter sections. Each can be considered as made up from half-sections.

Chebycheff) respectively. The differences are that the Butterworth response shows very little variation of attenuation in the pass-band, but the transition between pass-band and stop-band is not particularly steep. The Tchebycheff response shows more variation within the pass-band, but has a steeper slope between the bands (Figure 6.11). There is no difference in circuitry needed to provide the responses of these filters – the same filter circuit can be made to provide either form of response depending on the values of the components.

The two basic arrangements of components for a ladder network are T and Pi, illustrated in Figure 6.12. Either of these can be considered as being made out of half-sections, as illustrated, so that the construction of each half-section is the main factor in the design of a filter. Using inductors and capacitors, the half-section consists of a series and a shunt element, two reactive elements, so that the simplest type of filter is a two element type, but most practical filters use more than two elements in the half-section so as to achieve better performance.

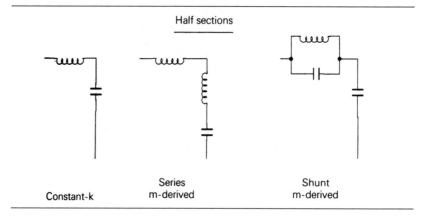

Figure 6.13   Examples of the three main types of low-pass filters.

Figure 6.13 shows the simplest low-pass filters, called the constant-k, series m-derived and shunt m-derived types. The k refers to the geometric mean impedance, the root of the product of input and output impedances for the half-section, and m refers to the quantity:

$$\sqrt{1 - \frac{f_c^2}{f_\infty^2}}$$

where f_c, is the cut-off frequency and f_∞ is the frequency of peak attenuation.

Even for these comparatively simple elements, the prediction of response is anything but easy, because each half-section is supposed to be connected to an impedance which matches its own (variable) impedance. In addition, expressions for the filter behaviour are often based on idealized lossless inductors and capacitors. This makes it all the more difficult to draw out the behaviour for a ladder network of several sections (Figure 6.14), in which the elements will inevitably interact.

The response curves for full sections of low-pass filters are illustrated in Figure 6.15, showing idealized attenuation and phase characteristics in terms of cut-off frequency and peak attenuation frequency. Note that the constant-k type of filter has its frequency of peak attenuation at a theoretically infinite frequency, whereas the m-derived types have an attenuation that reduces at frequencies beyond the peak. Figure 6.16 shows high-pass filter sections, once again using the constant-k, series m-derived and shunt m-derived

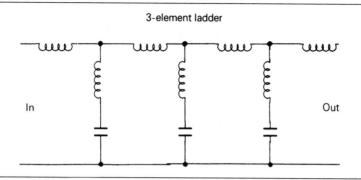

3-element ladder

Figure 6.14 An example of a complete ladder network, for which calculations are lengthy and best done by computer.

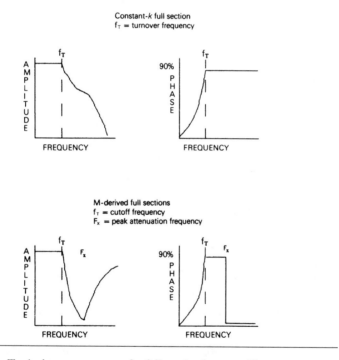

Constant-*k* full section
f_T = turnover frequency

M-derived full sections
f_T = cutoff frequency
F_x = peak attenuation frequency

Figure 6.15 Typical response curves for full-section low-pass filters.

forms. The performance curves for full sections are illustrated in Figure 6.17.

Band-pass and band-stop filters will usually incorporate more elements than a high-pass or low-pass type, and Figure 6.18 shows a

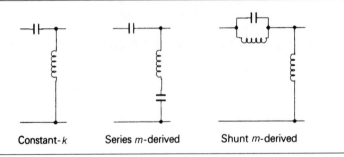

Constant-*k*   Series *m*-derived   Shunt *m*-derived

Figure 6.16   High-pass filter sections of the three main types.

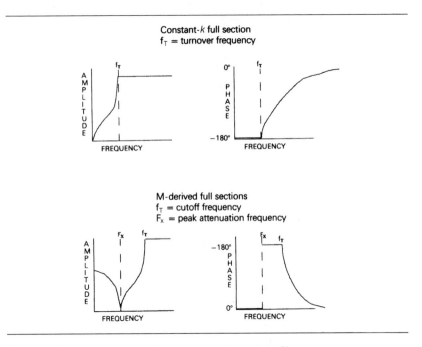

Figure 6.17   Performance of full sections of high-pass filters.

five element series type of band-pass filter along with the full-section responses. Filters of this complexity are seldom nowadays constructed with passive components because the behaviour of active filters and of the SAW type is so much more predictable and so much closer to ideal performance.

Two important filter types make use of resistor and capacitor networks. Figure 6.19 shows the twin-T network (a), with typical

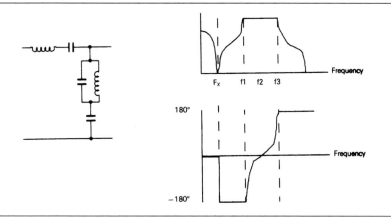

Figure 6.18   An example of a five element band-pass filter, and performance.

response curves for amplitude and phase (b), and Figure 6.20(a) shows the Wien bridge with its frequency and phase response curves in (b). Both examples have used as frequency determining components values of 10 nF and 10 kΩ, but note that the twin-T network requires one capacitor of double the 10 nF value and one resistor of half the 10 kΩ value. These two networks have been extensively used in oscillator circuits and in 'notch' filters, but generally in the frequencies below 100 kHz.

A delay line is a form of low-pass filter whose phase characteristics are more important than the attenuation characteristics. The aim is to cause the time delay of a signal (often a pulse or other non-sinusoidal signal) by times of up to a microsecond or so. Since the delay line is a low-pass filter, it will cause unavoidable attenuation, so that delays of more than about a microsecond involve so many sections of line that the attenuation is too large for practical use. For longer delays, transducer delay lines of the acoustic type (see Chapter 8) are preferred.

Suppressors are a special case of filters of the band-pass type, in which some variation of the attenuation in the pass-band can be accepted in exchange for high attenuation in the stop-band. The pass-band is often wide, covering the radio and/or TV range of frequencies, with high attenuation particularly at the lower frequencies. The design methods are as for filters generally, but some forms of suppression use non-linear resistors in addition to inductors and capacitors.

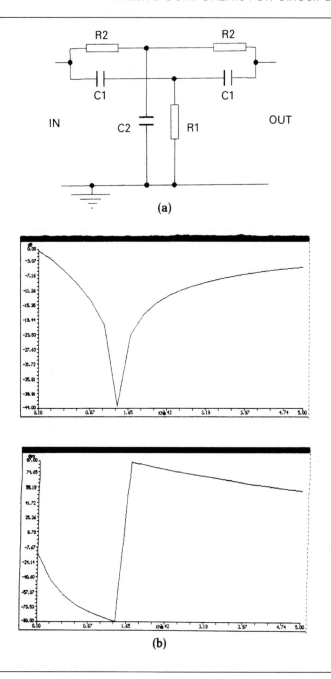

Figure 6.19 The twin-T notch filter circuit (a), with its frequency and phase response graphs (b). The values used are R1 = 5 kΩ, R2 = 10 kΩ, C1, C2 = 10 nF.

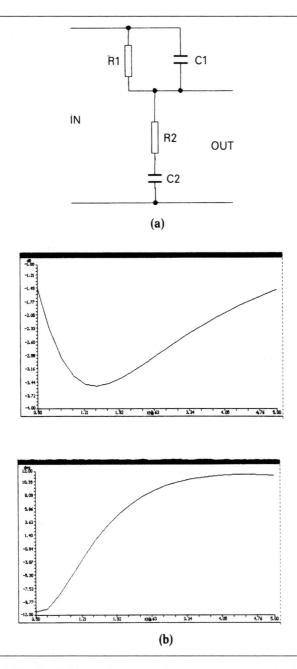

Figure 6.20 The Wein bridge notch filter circuit (a) with its frequency and phase response graphs (b). The values used are R1, R2 = 10 kΩ and C1, C2 = 10 nF.

Chapter 7

Electrochemical cells

Chemical cells and batteries

Chemical cells were the original source of DC, and have always been an important form of power supply for electronic equipment. Historically, cells and batteries have been in use for over two hundred years, and the problems that are encountered with one of the simplest and oldest types of cell are a good introduction to the reasons why so many diverse battery types exist nowadays, and to the technology that is used. Strictly speaking, a battery is an assembly of single cells, so that the action of a cell is the subject of this chapter.

Any type of chemical cell depends on a chemical action which is usually between a solid (the cathode plate) and a liquid, the electrolyte. The use of liquids makes cells less portable, and the trend for many years has been to use jellified liquids, and also materials that are not strong acids or alkalis. The voltage that is obtained from any cell depends on the amount of energy liberated in the chemical reaction, but only a limited number of chemical reactions can be used in this way, and for most of them, the energy that is liberated corresponds to a voltage of between 0.8 and 2.3 V per cell with one notable exception, the lithium cell. This range of voltage represents a fundamental chemical action that cannot be circumvented by refining the mechanical or electrical design of the cell.

The current that can be obtained from a cell is, by contrast, determined by the area of the conducting plates and the resistance of the electrolyte material, so that there is a relationship between physical size and current capability. The limit to this is purely practical, because if the cell is being used for a portable piece of equipment, a very large cell makes the equipment less portable and therefore less useful. Hundreds of types of cells have been invented and constructed since 1790, and most of them have been forgotten, even from school textbooks (although the Weston Standard Cell still occupies a place despite the fact that the more practical mercury button cell, bought from the local chemist at a tenth of the price, provides as useful a reference voltage). By the middle of the twentieth century, only one type of cell was commonly available, the Leclanché cell which is the familiar type of 'ordinary' torch cell. The introduction of semiconductor electronics, however, has revolutionized the cell and battery industry, and the requirements for specialized cells to use in situations calling for high current, long shelf life or miniature construction have resulted in the development and construction of cells from materials that would have been considered decidedly exotic in the earlier part of the century.

Primary and secondary cells

A primary cell is one in which the chemical reaction is not reversible. Once the cell is exhausted, because the electrolyte has dissolved all of the cathode material or because some other chemical (such as the depolarizer, see later) is exhausted, then recharging to the original state of the cell is impossible, although for some types of primary cell, a very limited extension of life can be achieved by careful recharging. In general, attempts to recharge a primary cell will usually result in the internal liberation of gases which will eventually burst explosively through the case of the cell. A secondary cell is one in which the chemical reaction is one that is reversible. Without getting into too much detail about what exactly constituted reversibility, reversible chemical reactions are not particularly common, and it is much more rarely that such a reaction can be

used to construct a cell, so that there is not the large range of cells of
the secondary type such as exists for primary cells. Even the nickel–
cadmium secondary cell which is used so extensively nowadays in
the form of rechargeable batteries is a development of an old
design, the nickel–iron cell due to Edison in the latter years of the
nineteenth century.

There is a third type of cell, the fuel cell, which despite very great
research efforts for some thirty years has not become as common as
was originally predicted. A fuel cell uses for its power a chemical
reaction which is normally combustion, the burning of a substance,
and is dealt with briefly in Chapter 6.

BATTERY CONNECTIONS

When a set of cells is connected together, the result is a battery. The
cells that form a battery could be connected in series, in parallel, or
in any of the series–parallel arrangements, but in practice the connection is nearly always in series. The effect of both series and parallel connection can be seen in Figure 7.1. When the cells are
connected in series, the open-circuit voltages (EMFs) add, and so
do the internal resistance values, so that the overall voltage is
greater, but the current capability is the same as that of a single cell.

When the cells are connected in parallel, the voltage is as for one
cell, but the internal resistance is much lower, because it is the resul-
tant of several internal resistances in parallel. This allows much
larger currents to be drawn, but unless the cells each produce
exactly the same EMF value, there is a risk that current will flow
between cells, causing local overheating. For this reason, primary
cells are never used connected in parallel, and even secondary cells,
which are more able to deliver and to take local charging current,
are seldom connected in this way except for recharging.

Higher currents are obtained by making primary cells in a variety
of sizes, with the larger cells being able to provide more current,
and having a longer life because of the greater quantity of essential
chemicals. The limit to size is portability, because if a primary cell
is not portable it has a limited range of applications. Secondary
cells have much lower internal resistance values, so that if high
current capability is required along with small volume, a secondary

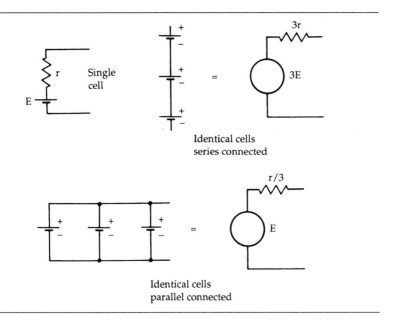

Figure 7.1 Connecting cells in series and in parallel.

cell is always used in preference to a primary cell. One disadvantage of the usual type of nickel–cadmium secondary cell in this respect, however, is a short 'shelf life', so that if equipment is likely to stand for a long time between periods of use, secondary cells may not be entirely suitable, because they will always need to be recharged just before use.

The important parameters for any type of cell are its open-circuit voltage (the EMF), its 'typical' internal resistance value, its shelf life, active life and energy content. The shelf life indicates how long a cell can be stored, usually at a temperature not exceeding $25°C$, before the amount of internal chemical action seriously decreases the useful life. The active life is less easy to define, because it depends on the current drain, and it is usual to quote several figures of active life for various average current drain values. The energy content is defined as EMF × current × active life, and will usually be calculated from the most favourable product of current and time. The energy content is more affected by the type of chemical reaction and the weight of the active materials than by details of design.

Cell origins

All of the cells that are used today can trace their origins to the voltaic pile that was invented by Alessandro Volta around 1782. Each portion of this device was a sandwich of cloth soaked in brine, Figure 7.2(a) and laid between one plate of copper and one plate of zinc. When sufficient of the sandwich cells were assembled into a battery, the voltage was enough to cause effects such as the heating of a thin wire, or the twitching of the leg of a (dead) frog – the effect discovered by Luigi Galvani. The sort of electrical systems we get in Italian cars came much later!

The next step was the simple cell, as we now call it, which used the metal zinc (the cathode) and the liquid sulphuric acid to provide the chemical reaction, and the other contact, the anode, that was needed was provided by a copper plate which was also dipped into the acid, Figure 7.2(b). The action is that when the zinc dissolves in the acid, electrons are liberated. These electrons can flow along a wire connected to the zinc, and back into the chemical system through the copper plate, so meeting the requirement for a closed path for electrons.

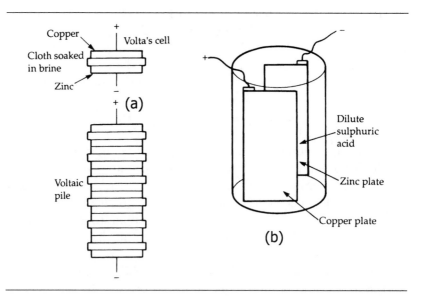

Figure 7.2 (a) Volta's original cell principle, and the battery (pile) derived from it. (b) The original form of wet simple cell.

In terms of conventional current flow, a decision made long before the existence of electrons was suspected, this is a current flowing from the positive copper plate, the anode, to the negative zinc plate, the cathode. All cells conform to this pattern of a metal dissolving in an acid or alkaline solution and releasing electrons that return to the cell by way of an inert conductor which is also immersed in the solution. The original zinc–sulphuric acid type of cell is known as the simple cell to distinguish it from the many types that have followed.

The simple cell has several drawbacks that make it unsuitable for use other than as a demonstration of principles. The use of sulphuric acid in liquid form makes the cell unsuitable for any kind of portable use, since acid can spill and even at the dilution used for the simple cell it can cause considerable damage. The cell cannot be sealed, because as the zinc dissolves it liberates hydrogen gas which must be vented.

There are more serious problems. The sulphuric acid will dissolve the zinc, although at a slower rate, even when no circuit exists, so that the cell has a very short shelf life and not much active life. In addition, the voltage of the cell, which starts at about 1.5 V, rapidly decreases to zero when even only a small current is taken because the internal resistance rises to a large value as the cell is used. This makes the cell unusable until the zinc is removed, washed, and then reinserted. No doubt if Alessandro Volta had been working with a government grant rather than on private funds, he would have been told that his experiments were leading nowhere and would no longer be funded (and what use, pray, could ever be found for such an invention?).

The efforts that were made to understand the faults of the simple cell have led to the development of considerably better cells, because by understanding principles we are better able to design new products. The problem of the zinc dissolving even with no circuit connected was solved by using very pure zinc or by coating the zinc with mercury. The problem is one of *local action*, meaning that the impurities in the zinc act like anodes, forming small cells that are already short circuited. By using very pure zinc, this local action is very greatly reduced, but in the eighteenth century purification of metals had not reached the state that we can expect nowadays. Mercury acts to block off the impurities without itself acting as an anode, and this was a much easier method to use at the time.

The rapid increase in internal resistance proved to be a more difficult problem, and one that could not be solved other than by redesigning the cell. Curiously enough, however, the problem of the increasing internal resistance was later used as a way of constructing electrolytic capacitors, see Chapter 1. The problem is that dissolving zinc in sulphuric acid releases hydrogen gas, and this gas coats the surface of the anode as it is formed, an action that was originally called polarization. The gas appears at the anode because of the action of the electrons entering the solution from the external circuit. Because hydrogen is an insulator, the area of the anode that can be in electrical contact with the sulphuric acid is greatly reduced by this action, so that the internal resistance increases. When local action is present, the internal resistance will increase from the moment that the cell is assembled, although for the pure-zinc cell or the type in which the zinc has been coated (amalgamated) with mercury, the internal resistance increases only while the cell is used.

The problem can be solved only by removing the hydrogen as it forms or by using a chemical reaction that does not generate any gas, and these are the solutions that have been adopted by every successful cell type developed since the days of Volta. The removal of hydrogen is achieved by using an oxidizing material, the depolarizer, which has to be packed around the anode. The depolarizer must be some material that will not have any chemical side-effects, and insoluble materials like manganese (II) oxide have been used very successfully in the past and are still widely used.

The Leclanché cell

The cell that was developed by the French chemist Leclanché in the nineteenth century has had a remarkably long history, and in its 'dry' form is still in use, although now grandified by the title of *carbon–zinc cell*. In its original form, Figure 7.3, the electrolyte was a liquid, a solution of ammonium chloride. This is mildly acid, but not fiercely corrosive in the way that sulphuric acid is, and one consequence of using this less acidic electrolyte is that the zinc, even if not particularly pure, does not dissolve in the solution to the same extent when no current is passing in the external circuit. Local

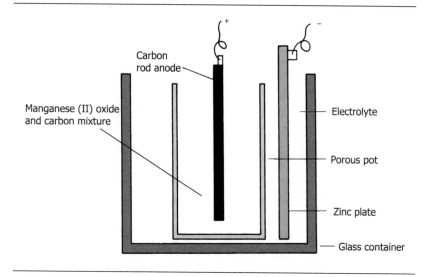

Carbon rod anode

Manganese (II) oxide and carbon mixture

Electrolyte

Porous pot

Zinc plate

Glass container

Figure 7.3 The original form of the Leclanché wet cell.

action is still present, but greatly reduced as compared to a zinc–acid type of cell. The anode for the cell is a rod of carbon, a material that is chemically inert and therefore unattacked by the electrolyte. The carbon rod is surrounded by a paste of manganese dioxide, all contained inside a porous pot so that the electrolyte keeps the whole lot wet and conducting. The action when current flows is that zinc dissolves in the mildly acid solution, releasing electrons which then travel through the circuit.

At the anode, the electrons would normally react with the water in the liquid to produce hydrogen, but the action of the manganese dioxide is to absorb electrons in preference to allowing the reaction with the water to proceed, producing a different oxide of manganese (a reduced state). As the cell operates, the zinc is consumed, as also is the manganese dioxide, and when either is exhausted the cell fails. The open-circuit voltage is about 1.5 V, and the internal resistance can be less than one ohm. The older form of the Leclanché cell was in service for operating doorbells and room indicators from mid-Victorian times, and some that had been installed in these days were still working in the late 1930s.

The reason is that the Leclanché cell was quite remarkably renewable. The users could buy spare zinc plates, spare ammonium chloride (which could also be used for smelling salts) and spare

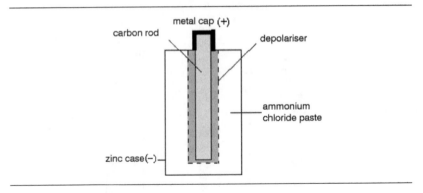

Figure 7.4   The modern form of dry carbon–zinc cell.

manganese dioxide, so that the cell could be given an almost indefinite life on the type of intermittent use that it had. Some worked for well over twenty years without any attention at all, tucked away in a cupboard on a high shelf.

The 'dry' form of the Leclanché cell is the type that until quite recently was the only familiar form of primary cell. The construction, Figure 7.4, follows the principles of the older wet type of cell, but the ammonium chloride electrolyte is in jelly form rather than liquid, and the manganese oxide is mixed with graphite and with some of the jelly to keep it also moist and conducting. The action is the same, but because the dry cell is usually smaller than the wet variety and because its jelly electrolyte is less conductive, this form of the cell has generally a higher internal resistance than the old wet variety. The advantage of portability, however, totally overrules any disadvantages of higher internal resistance, making this the standard dry cell for most of the twentieth century.

The carbon–zinc dry cell, as it is more often called now, fails totally either when the zinc is perforated or when the manganese dioxide is exhausted. One of the weaknesses of the original design is that the zinc forms the casing for the cell, so that when the zinc becomes perforated, the electrolyte can leak out, and countless users of dry cells will have had the experience of opening a torch or a transistor radio battery compartment to find the usual sticky mess left by leaking cells. The term 'dry' cell never seems quite appropriate in these circumstances. The problem cannot simply be dealt with by using a thicker zinc casing and by restricting the amount of manganese dioxide so that the cell fails because of high internal resistance before the zinc is used up.

The carbon–zinc cell does not have a particularly long shelf life and once it has been used, the electrolyte starts to dissolve the zinc at a slow but inexorable rate. This corresponds to an internal current within the cell, called the self-discharge current. Perforation will therefore invariably occur when an exhausted cell is left inside equipment, and the higher the temperature at which the cell is kept, the faster is the rate of attack of the zinc.

This led to the development of leakproof cells with a steel liner surrounding the zinc. Leakproofing in this way allowed a much thinner zinc shell to be used, so cutting the cost of the cell (although it could be sold at a higher price because of the leakproofing) and allowing the cell to be used until a much greater amount of the zinc had been dissolved. Leakproofing is not foolproof, and even the steel shell can be perforated in the course of time, or the seals can fail and allow electrolyte to spill out. Nevertheless, the use of the steel liner has considerably improved the life of battery-operated equipment.

The alkaline primary cells

A different group of cell types makes use of alkaline rather than acid electrolytes, so that although the principle of a metal dissolving in a solution and releasing electrons still holds good, the detailed chemistry of the reaction is quite different. On the assumption that the reader of this book will be considerably more interested in the electrical characteristics of these cells rather than the chemistry, we will ignore the chemical reactions unless there is something about them that requires special notice. One point that does merit attention is that the alkaline reactions do not generate gas, and this allows the cells to be much more thoroughly sealed than the zinc–carbon type. It also eliminates the type of problems that require the need of a depolarizer, so that the structure of alkaline cells can, in theory at least, be simpler than that of the older type of cell. Any attempt to recharge these cells other than by well-designed (microprocessor-controlled) circuitry will generate gas and the pressure will build up until the container fractures explosively.

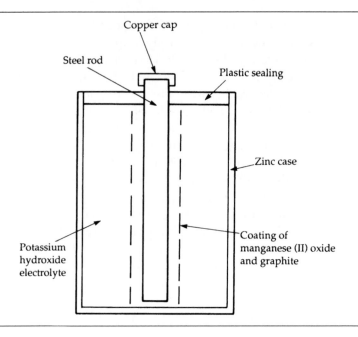

Figure 7.5   Typical cross-section of a manganese–alkaline cell.

The best-known alkaline type of cell is the manganese–alkaline, whose construction is illustrated in Figure 7.5. This was invented by Sam Ruben in the USA in 1939 and was used experimentally in some wartime equipment, but the full-scale production of manganese–alkaline cells did not start until the 1960s. The cell uses zinc as the cathode, with an electrolyte of potassium hydroxide solution, either as liquid or as jelly, and the anode is a coating of manganese (II) oxide mixed with graphite and laid on steel. The cell is sealed because the reaction does not liberate gas, and the manganese (II) oxide is being used for its manganese content rather than for its oxygen content as a depolarizer.

The EMF of a fresh cell is 1.5 V, and the initial EMF is maintained almost unchanged for practically the whole of the life of the cell. The energy content, weight for weight, is higher than that of the carbon–zinc cell by a factor of 5–10, and the shelf life is very much better due to an almost complete lack of secondary action. All of this makes these cells very suitable for electronics use, particularly for equipment that has fairly long inactive periods followed by large current demand. Incidentally, although the cells use alkali rather than acid, potassium hydroxide is a caustic material which will dissolve the skin and is extremely dangerous to the eyes. An alkaline cell must never be opened, nor should any attempt ever be made to recharge it other than with specialized charging equipment.

Miniature (button) cells

The miniature cells are the types specified for deaf-aids, calculators, cameras and watches, but they are quite often found in other applications, such as for backup of memory in computing applications and for 'smart-card' units in which a credit card is equipped with a complete microprocessor and memory structure so that it keeps track of transactions. The main miniature cells are silver oxide and mercury, but the term mercury cell can be misleading, because metallic mercury is not involved.

The mercuric oxide button cell, to give it the correct title, uses an electrolyte of potassium hydroxide (Figure 7.6) which has had zinc oxide dissolved in it until saturated, so that the cell can be classed as an alkaline type. The cathode is the familiar zinc, using either a

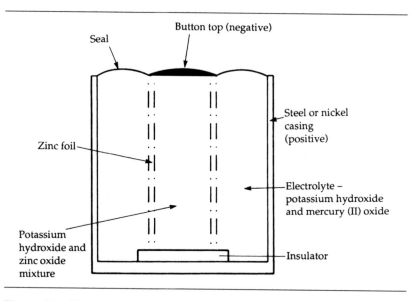

Figure 7.6 The construction of a mercuric oxide cell.

cylinder of perforated zinc foil or a sintered zinc-powder cylinder fastened to the button-top of the cell and insulated from the bottom casing. The anode is a coating of mercury (I) oxide mixed with graphite to improve conductivity and coated on nickel-plated steel or stainless steel which forms the casing of the cell. The EMF of such cells is low, 1.2–1.3 V, and the energy content is high, with long shelf life due to the absence of local action.

The silver oxide cell is constructed in very much the same way as the mercuric oxide cell, but using silver (I) oxide mixed with graphite as the anode. The cathode is zinc and the electrolyte is potassium hydroxide as for the mercuric oxide cell. The EMF is 1.5 V, a value that is maintained at a steady level for most of the long life of the cell. The energy content is high and the shelf life long.

All of these miniature cells are intended for very low current applications, so that great care should be taken to avoid accidental discharge paths. If the cells are touched by hand, this will leave a film of perspiration that is sufficiently conductive to shorten the life of the cell drastically. When these cells are fitted, they should be moved and fitted with tweezers, preferably plastic tweezers or with dry rubber gloves if you need to use your hands. These cells should not be recharged, nor disposed of in a fire. The mercury type is particularly hazardous if mercury compounds are released, and they should be returned to the manufacturer for correct disposal if this is possible, otherwise they should be disposed of by a firm that is competent to handle mercury compounds.

Lithium cells

Lithium is a metal akin to potassium and sodium which is highly reactive, so much so that it cannot be exposed to air and reacts with explosive violence with water. The reactive nature of lithium metal means that a water solution cannot be used as the electrolyte and much research has gone into finding liquids which ionize to some extent but which do not react excessively with lithium. A sulphur–chlorine compound, thionyl chloride, is used, with enough dissolved lithium salts to make the amount of ionization sufficient for the conductivity that is needed. The lithium cell is the most recently developed type of cell, remarkable for its unexpectedly high EMF.

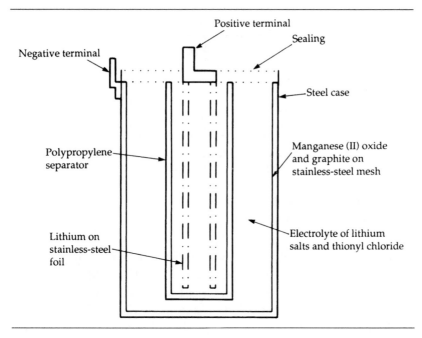

Figure 7.7   The construction of a lithium cell.

The lithium, Figure 7.7, is coated on to a stainless-steel mesh which is separated from the rest of the cell by a porous polypropylene container. The anode is a mixture of manganese (II) oxide and graphite, also coated on to stainless-steel mesh. The whole cell is very carefully sealed. The reaction can be used to provide a cell with an exceptionally high EMF of 3.7 V, very long shelf life of ten years or more, and high energy content. The EMF is almost constant over the life of the cell, and the internal resistance can be low.

Lithium cells are expensive, but their unique characteristics have led to them being used in automatic cameras where focusing, film wind, shutter action, exposure and flash are all dependent on one battery, usually a two-cell lithium type. For electronics applications, lithium cells are used mainly for memory backup, and very often the life of the battery is as great as the expected lifetime of the memory itself. The cells are sealed, but since excessive current drain can cause a build-up of hydrogen gas, a 'safety valve' is incorporated in the form of a thin section of container wall which will blow out in the event of excess pressure. Since this will allow the

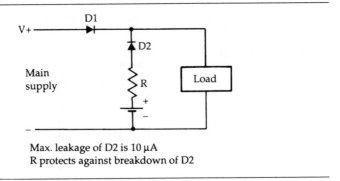

Max. leakage of D2 is 10 μA
R protects against breakdown of D2

Figure 7.8 A recommended reverse-current protection circuit for a lithium cell in a simple backup application.

atmosphere to reach the lithium, with risk of fire, the cells should be protected from accidental overcurrent, which would cause blow-out. A recommended protection circuit is illustrated in Figure 7.8.

This is for use in applications where the lithium cell is used as a backup, so that D1 conducts during normal memory operation and D2 conducts during backup. Short-circuit failure of D2 would cause the lithium cell to be charged by the normal supply, and the resistor R will then limit the current to an amount which the cell manufacturer deems to be safe. If the use of a resistor would cause too great a voltage drop in normal backup use, it could be replaced by a quick-blowing fuse, but this has the disadvantage that it would cause loss of memory when the main supply was switched off.

Lithium cells must never be connected in parallel, and even series connection is discouraged and limited to a maximum of two cells. The cells are designed for low load currents, and Figure 7.9 shows a typical plot of battery voltage, current and life at 20°C. Some varieties of lithium cells exhibit voltage lag, so that the full output voltage is available only after the cell has been on load for a short time – the effect becomes more noticeable as the cell ages. Another oddity is that the capacity of a lithium cell is slightly lower if the cell is not mounted with the +ve terminal uppermost.

Secondary cells

A secondary cell makes use of a reversible chemical process, so that

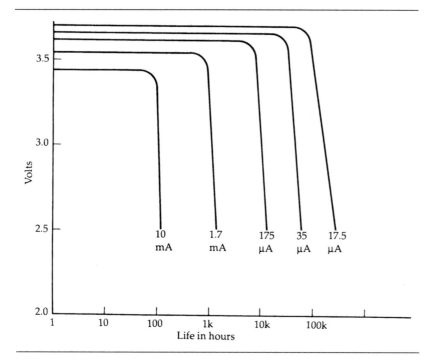

Figure 7.9 Typical graphs of lithium cell voltage, current and life at normal room temperature levels.

when the cell is discharged, reverse current into the cell will re-charge it by restoring the original chemical constitution. Unlike primary cell reactions, reversible reactions of this type are unusual and only two basic types are known, the lead–acid type and the alkali–metal type, both of which have been used for a considerable time.

The lead–acid cell construction principle is illustrated in Figure 7.10. Both plates are made from lead and are perforated to allow them to be packed with the active materials. One, the positive plate (anode), is packed with lead (IV) oxide, and the negative plate (cathode) is packed with spongy or sintered lead which has a large surface area. Both plates are immersed in sulphuric acid solution. The acidity is much greater than that of the electrolytes of any of the acidic dry cells, and very great care must be taken when working with lead–acid cells to avoid any spillage of acid or any charging fault that could cause the acid to boil or to burst out of the casing. In addition, the recharging of a vented lead–acid cell

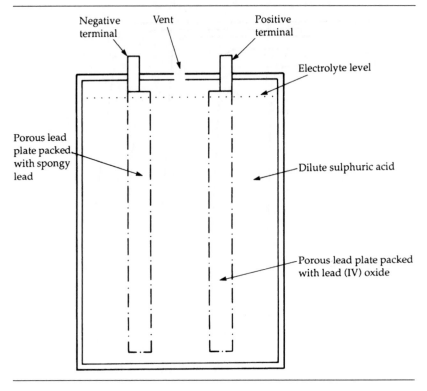

Figure 7.10 Principles of the lead-acid secondary cell.

releases hydrogen and oxygen as a highly explosive mixture which will detonate violently if there is any spark nearby.

The EMF is 2.2 V (nominally 2.0 V), and the variation in voltage is quite large as the cell discharges. Figure 7.11 shows typical discharge graphs for light-load and heavy-load respectively.

The older vented type of lead–acid cell is now a rare sight, and modern lead–acid cells are sealed, relying on better control of charging equipment to avoid excessive gas pressure. The dry type of cell uses electrolyte in jelly form so that these cells can be used in any operating position. Cells that use a liquid electrolyte are constructed with porous separator material between the plates so that the electrolyte is absorbed in the separator material, and this allows these cells also to be placed in any operating position. Since gas pressure build-up is still possible if charging circuits fail, cells are equipped with a pressure-operated vent that will reseal when pressure drops again.

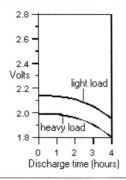

Figure 7.11 Voltage drop on discharging a lead–acid cell.

Lead–acid cells are used in electronics applications mainly as backup power supplies, as part of uninterruptible power systems, where their large capacities and low internal resistance can be utilized. Capacity is measured in ampere-hours, and sizes of 9 Ah to 110 Ah are commonly used. Care should be taken in selecting suitable types – some types of lead–acid cells will self-discharge considerably faster than others and are better suited to applications where there is a fairly regular charge/discharge cycle than for backup systems in which the battery may be used only on exceptional occasions and charging is also infrequent. Figure 7.12 shows the self-discharge rates of jelly-electrolyte cells at various temperatures, taking the arbitrary figure of 50% capacity as the discharge point.

Lead–acid batteries need to be charged from a constant-voltage source of about 2.3 V per cell at 20°C – Figure 7.13 shows the variation of charging voltage per cell with ambient temperature of the cell. Cells can be connected in series for charging provided that all of the cells are of the same type and equally discharged. A suitable multi-cell charger circuit is illustrated in Figure 7.14, courtesy of RS Components.

For batteries of more than 24 V (12 cells) the charging should be in 24 V blocks, or a charging system used that will distribute charging so that no single cell is being overcharged. Parallel charging can be used if the charger can provide enough current. The operating life of a lead–acid cell is usually measured in terms of the number of charge/discharge cycles, and is greater when the cell is used with fairly high discharge currents – the worst operating

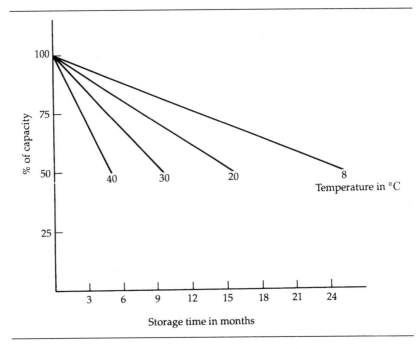

Figure 7.12   Self-discharge graphs for a jelly type of lead–acid cell.

conditions are of slow discharge and erratic recharge intervals – the conditions that usually prevail when these cells are used for backup purposes.

One condition to avoid is deep discharge, when the cell has been left either on load or discharged for a long period. In this state, the terminal voltage falls to 1.6 V or less and the cell is likely to be permanently damaged unless it is immediately recharged at a very low current over a long period. Typical life expectancy for a correctly operated cell is of the order of 750–6000 charge/discharge cycles.

NICKEL–CADMIUM CELLS

The original type of alkaline secondary cell, invented by Edison at the turn of the century, was the nickel-cathode iron-anode type, using sodium hydroxide as the electrolyte. The EMF is only 1.2 V, but the cell can be left discharged for long periods without harm,

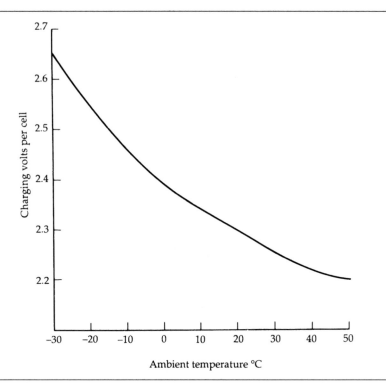

Figure 7.13 Temperature variation of charging voltage for a lead–acid cell.

and will withstand much heavier charge and discharge cycles than the lead–acid type. Although the nickel–iron alkaline secondary cell still exists powering milk–floats and fork-lift trucks, it is not used in the smaller sizes because of the superior performance of the nickel–cadmium type of cell which is now the most common type of secondary cell used for cordless appliances and in electronics uses.

Nickel–cadmium cells can be obtained in two main forms, mass-plate and sintered plate. The mass-plate type used nickel and cadmium plates made from smooth sheet, the sintered type has plates formed by moulding powdered metal at high temperatures and pressures, making the plates very porous and of much greater surface area. This makes the internal resistance of sintered-plate cells much lower, so that larger discharge currents can be achieved. The mass-plate type, however, have much lower self-discharge rates and are more suitable for applications in which recharging is

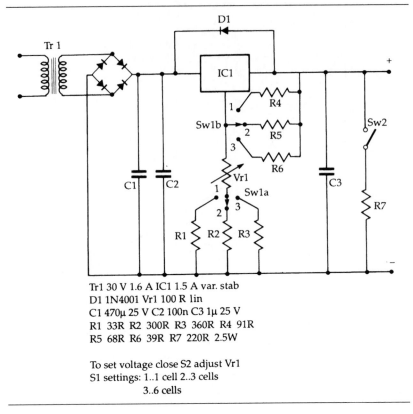

Tr1 30 V 1.6 A IC1 1.5 A var. stab
D1 1N4001 Vr1 100 R lin
C1 470μ 25 V C2 100n C3 1μ 25 V
R1 33R R2 300R R3 360R R4 91R
R5 68R R6 39R R7 220R 2.5W

To set voltage close S2 adjust Vr1
S1 settings: 1..1 cell 2..3 cells
               3..6 cells

Figure 7.14 A circuit for a multi-cell charger, due to RS Components.

not frequent. Typical life expectancy is from 700 to 1000 charge/dis-
charge cycles.

One very considerable advantage of the nickel–cadmium cell is
that it can be stored for 5 years or more without deterioration.
Although charge will be lost, there is nothing corresponding to the
deep discharge state of lead–acid cells that would cause irreversible
damage. The only problem that can lead to cell destruction is
reverse polarity charging. The cells can be used and charged in any
position, and are usually supplied virtually discharged so that they
must be fully charged before use. Most nickel–cadmium cell types
have a fairly high self-discharge rate, and a cell will on occasions
refuse to accept charge until it has been 'reformed' with a brief
pulse of high current. Cells are usually sealed but provided with a
safety vent in case of incorrect charging.

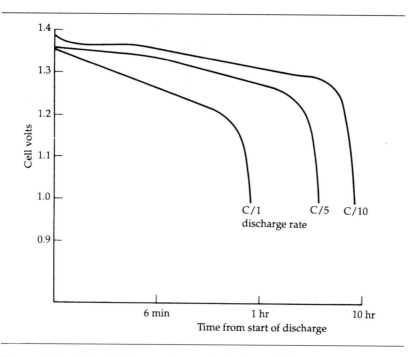

Figure 7.15   Typical discharge characteristics for small nickel–cadmium cells.

In use, the nickel–cadmium cell has a maximum EMF of about 1.4 V, 1.2 V nominal, and this EMF of 1.2 V is sustained for most of the discharge time. The time for discharge is usually taken arbitrarily as the time to reach an EMF of 1 V per cell, and Figure 7.15 shows typical voltage–time plots for a variety of discharge rates. These rates are noted in terms of capacity, ranging from one-fifth of capacity to five times capacity, when capacity is in ampere-hours and discharge current in amps. For example, if the capacity is 10 Ah, then a C/5 discharge rate means that the discharge current is 2 A.

Charging of nickel–cadmium cells must be done from a constant-current source, in contrast to the constant-voltage charging of lead–acid types. The normal rate of charge is about one-tenth of the Ah rate, so that for a 20 Ah cell, the charge rate would be 2 A. Sintered types can be recharged at faster rates than the mass-plate type, but the mass-plate type can be kept on continuous trickle charge of about 0.01 of capacity (for example, 10 mA for a cell of 1 Ah capacity). At this rate, the cells can be maintained on charge

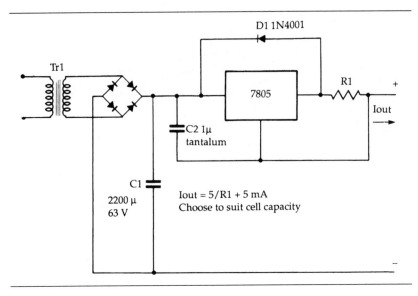

Figure 7.16 A recommended charging circuit for nickel–cadmium cells, courtesy of RS Components.

for an extended period after they are fully charged, but this overcharge period is about three times the normal charging time. Equipment such as portable and cordless phones which would otherwise be left on charge over extended intervals such as bank holiday weekends and office holidays should be disconnected from the charger rather than left to trickle charge. This means that a full charge will usually be needed when work resumes, but the life of the cells can be considerably extended if the very long idle periods of charging can be avoided. Another option is to leave the equipment switched on so as to discharge the cells, and fit the mains supply with a timer so that there will periodic recharging.

Figure 7.16 shows a recommended circuit for recharging, courtesy of RS Components. This uses a 7805 regulator to provide a fixed voltage of 5 V across a resistor, so that the value of the current depends on the choice of resistor and not on the voltage of the cell. The value of the resistor has to be chosen to suit the type of cell being recharged; values from 10 Ω to 470 Ω are used depending on the capacity of the cell. Because the regulator system is floating with respect to ground, this can be used for charging single cells or series sets of a few cells. Ready-made chargers are also available which will take various cells singly or in combination, with the correct current regulation for each type of cell.

- A form of silver cell has also been used in rechargeable form. This uses an anode of porous zinc, usually a sintered component, with a silver (I) oxide and graphite cathode. The electrolyte is potassium hydroxide solution that has been saturated with zinc hydroxide. The cell can take a limited number of recharging cycles, but is now uncommon.

Chapter 8

Transducing components

A transducer is a device that converts energy from one form into another and in the broadest sense even a resistor qualifies for this description since it converts electrical energy into heat energy. We normally reserve the term of transducer, however, for the components in which the conversion is desired and used, and in this book we are concerned only with passive transducers that require no energy supply other than the energy that is being converted. Many of these devices are specialized, and only a broad outline of some of the most common devices will be dealt with in this book for reasons of space. The reader should consult a more specialized text for further information.

The main features of transducers, as far as their status as passive components is concerned, are their sensitivity or efficiency, their electrical impedance, their linearity and their bandwidth or frequency response. Sensitivity or efficiency refers to the fraction of input energy which is converted to output energy, and for many transducers this can be very low, of the order of a few per cent. The percentage efficiency is often masked by quoting the sensitivity in terms of the amount of output per unit input. Since the units of input and output will not necessarily be the same, the efficiency of conversion is not obvious.

The electrical impedance of transducers can range from a fraction of an ohm to many megohms, and it affects the type of connections that can be made to the transducer. In addition, if the transducer has an electrical output, the impedance level usually goes hand in hand with the typical electrical output level. Linearity means the extent to which the output is proportional to the input, so that a perfectly linear device would have a graph of output plotted against input that would be a straight line.

Some of the earliest transducers to be used in electronics were the electro-acoustical type, microphones and loudspeakers. A sound wave is the waveform caused by a vibration which will, in turn, cause an identical vibration to be set up in any material affected by the sound wave. The transducer for sound energy to electrical energy is the microphone, and microphone types are classified by the type of transducer they use. The characteristics of a microphone are both acoustic and electrical. The overall sensitivity is expressed as millivolts or microvolts of electrical output per unit intensity of sound wave, or in terms of the acceleration produced by the sound wave. In addition, though, the impedance of the microphone is of considerable importance. A microphone with high impedance usually has a fairly high electrical output, but the high impedance makes it very susceptible to hum pickup, either magnetically or electrostatically coupled. A low impedance is usually associated with very low output, but, provided that the layout and shielding are both good, hum pickup is almost negligible.

Another important factor is whether the microphone is directional or omnidirectional. If the microphone operates by sensing the pressure of the sound wave, then the microphone will be omnidirectional, picking up sound arriving from any direction. If the microphone detects the velocity (speed and direction) of the sound wave, then it is a directional microphone, and the sensitivity has to be measured in terms of direction as well as amplitude of sound wave. The microphone types are known as pressure or velocity operated, omnidirectional or in some form of directional response (such as cardioid). The acoustic construction, rather than the type of transducer, determines whether a microphone is directional or not.

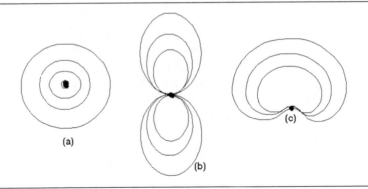

Figure 8.1 Simplified response curves for microphone types: (a) omnidirectional; (b) velocity-operated; (c) cardioid (heart-shaped).

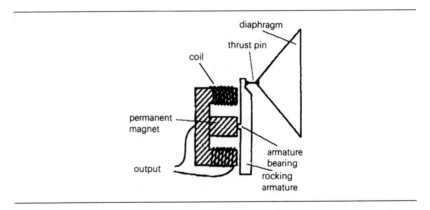

Figure 8.2 Moving-iron microphone principle. Movement of the diaphragm alters the magnetic circuit, inducing a voltage in the coil.

Microphone types

The principle of the moving-iron (variable reluctance) microphone is illustrated in Figure 8.2. A powerful magnet contains a soft-iron armature in its magnetic circuit, and this armature is attached to a diaphragm. The magnetic reluctance of the circuit alters as the armature moves, and this in turn alters the total magnetic flux in the magnetic circuit. A coil wound around the magnetic circuit at any point will give a voltage which is proportional to each change of magnetic flux, so that the electrical wave from the microphone is

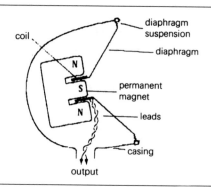

Figure 8.3   The moving-coil microphone principle. The movement of the coil in the magnetic field causes the output voltage.

proportional to the acceleration of the diaphragm. The linearity of the conversion can be reasonable for small amplitudes of movement of the armature, very poor for large amplitudes. Both the linearity, and the amplitude over which linearity remains acceptable can be improved by appropriate shaping of the armature and careful attention to its path of vibration. These features depend on the maintenance of close tolerances in the course of manufacturing the microphones, so that there will inevitably be differences in linearity between samples of microphones of this type from the same production line. The output level from a moving-iron microphone can be high, of the order of 50 mV, and the output impedance is fairly high, typically several hundred ohms. Magnetic shielding is always needed to reduce mains hum pickup in the magnetic circuit.

The moving-coil microphone uses a constant-flux magnetic circuit in which the electrical output is generated by moving a small coil of wire in the magnetic circuit (Figure 8.3). The coil is attached to a diaphragm, and as before the maximum output occurs as the coil reaches maximum velocity between the peaks of the sound wave so that the electrical output is at 90 degrees phase angle to the sound wave. The coil is usually small, and its range of movement very small, so that linearity is excellent, impedance is low and the signal output is also low.

The ribbon microphone is the logical conclusion of the moving-coil principle, in which the coil has been reduced to a strip of conducting ribbon (Figure 8.4), with the signal being taken from the ends of the ribbon. An intense magnetic field is used, so that the

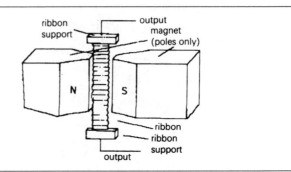

Figure 8.4 The ribbon microphone, the form of construction that provides the highest-quality of conversion from sound to electrical waves.

movement of the ribbon cuts across the maximum possible magnetic flux to generate an electrical output whose peak value is as usual at 90 degrees phase to the sound wave. The ribbon microphone is velocity operated, has excellent linearity and very low impedance. Its signal output is very low, however, so that a preamplifier usually has to be incorporated into the microphone.

The piezoelectric microphone uses a piezoelectric crystal element either alone or (not so commonly nowadays) connected to a diaphragm. The crystal, a material such as barium titanate, is one in which the arrangement of atoms ensures that any tiny displacement of the atoms due to vibration will cause a voltage to be generated across the crystal, and this is sensed by depositing conducting films across opposite faces of the crystal, to which the output leads are connected. The impedance level is of the order of several megohms, as distinct to a few ohms for a moving-coil type, and the output is in the high millivolt range rather than microvolts.

The capacitor microphone is a remarkable example of a principle that was comparatively neglected until another equally old idea was harnessed along with it. The outline of a capacitor microphone is illustrated in Figure 8.5. The amount of electrical charge between two surfaces is fixed and one of the surfaces is a diaphragm which can be vibrated by a sound wave. The vibration causes a variation of capacitance which, because of the fixed charge, causes in turn a voltage wave. The output impedance is very high, and the amount of output depends on the normal spacing between the plates – the smaller this spacing, the greater the output for a given amplitude of sound wave, and the poorer the linearity.

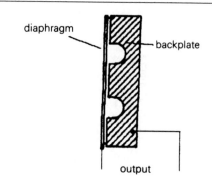

diaphragm

backplate

output

Figure 8.5 Capacitor microphone principle in its original form. This type of microphone has enjoyed a revival thanks to the use of electret materials.

A high-voltage supply (the polarizing voltage) is needed to provide the equivalent of a fixed charge by being connected across the plates through a resistor of very large value, and this polarizing requirement at one time deterred the widespread use of the capacitor microphone despite its good performance. The revival of the capacitor microphone came about as a result of improving the technology of an old idea, the electret.

An electret is the electrostatic equivalent of a magnet, a piece of insulating material which is permanently charged. A slab of electret is therefore the perfect basis for a capacitor microphone, providing the fixed charge that is required without the need for a polarizing voltage supply. This allows very simple construction of a capacitor microphone, consisting only of a slab of electret metallized on the back, a metal (or metallized plastic) diaphragm, and a spacer ring (Figure 8.6), with the connections taken to the conducting surface of the diaphragm and of the electret. This is now the type of microphone which is built into cassette recorders, and even in its simplest and cheapest versions is of considerably better audio quality than the piezoelectric types that it displaced. The use of a FET preamp solves the problem of the high impedance of this type of microphone.

Electrets currently in use consist of plastics materials such as Perspex which are subjected to an intense electric field while they are solidifying. There is no source of ready-made electrets, because the usual requirement for electrets calls for a custom shape of

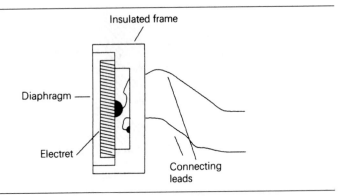

Figure 8.6 The electret microphone in cross-section. There is no need for a polarizing voltage, but the impedance of the microphone is very high, and a FET preamp must be used.

plastic, and it's more convenient for the end-user to form the plastic and polarize it.

Loudspeakers and earphones

Earphones have been in use for considerably longer than micro-phones or loudspeakers, since they were originally used for electric telegraphs. There is an earphone and loudspeaker form of construc-tion to match each type of microphone listed above, because these transducer types are all reversible (something quite rare among transducers). The task of the earphone is somewhat simpler than that of the loudspeaker, and the construction of an earphone that can provide acceptable quality of sound is very much simpler (and correspondingly cheaper) than that of a loudspeaker, since the earphone can use a small diaphragm, and ensure that the sound waves from this diaphragm are coupled directly to the ear cavity. The power that is required is in the low milliwatt level, and even a few milliwatts can produce considerable pressure amplitude at the eardrum – often more than is safe for the hearing.

A loudspeaker, by contrast, has its sound waves launched into a space whose properties are unknown, and it has to be housed in a cabinet whose resonances, dimensions and shape will considerably modify the performance of the loudspeaker unit. The assembly of loudspeaker and cabinet will be placed in a room whose dimensions and furnishing are outside the control of the loudspeaker designer, so that a whole new set of resonances and the presence of damping material must be considered. The transducer of a loudspeaker system is sometimes termed the 'pressure unit', and its task is to transform an electrical wave, which can be of a very complex shape, into an air-pressure wave of the same waveform. To do this, the device requires a motor unit, transforming electrical waves into vibration, and a diaphragm which will move sufficient air to make the effect audible. The diaphragm is one of the main problems of loudspeaker design, because it must be very stiff, curved, very light and free of resonances, an impossible combination of virtues. Practically every material known has been used for loudspeaker diaphragms at some time, from the classic varnished paper to titanium alloy and carbon fibre, and almost every shape variation on the traditional cone has been used.

The efficiency of loudspeakers is notoriously low, around 1%, mainly because of the acoustic impedance matching problem. In simple terms, most loudspeakers move a small amount of air with a comparatively large amplitude, whereas to produce a sound wave effectively they ought to move a very large amount of air at a comparatively low amplitude. This mismatch can be remedied to some extent by housing the loudspeaker in a suitable enclosure.

The efficiency is lowest at low frequencies, typically 20 Hz, and a loudspeaker will be designed so that the efficiency does not rise greatly above the low-frequency level. This greatly helps in achieving a reasonably level frequency response. A loudspeaker designed specifically for a small range of (higher) frequencies can be much more efficient, typically 10% or more, and some piezoelectric warning devices, as fitted to smoke alarms, can achieve almost 25% efficiency.

The first type of successful earphone transducer was a moving-iron type, and this principle was used extensively for both earphones and for loudspeakers earlier in the twentieth century. Moving-iron units are seldom encountered now except in telephone earpieces. The conventional telephone earphone uses a magnetized metal diaphragm so that the variation of magnetization of the fixed coil will ensure the correct movement of the diaphragm. The unit is sensitive but the linearity is poor and moving-iron units are seldom used.

The moving-coil principle as applied to loudspeakers and earphones has, by contrast, been widely adopted, and the vast majority of loudspeakers use this principle. The use of moving-coil earphones has been less common in the past, but these are now in widespread use thanks to the miniature cassette player vogue. As applied to earphones, moving-coil construction permits good linearity and controllable resonances, since the amount of vibration is very small and the moving-coil unit is light and can use a diaphragm of almost any suitable material. A variation of the moving-coil principle that has been successfully used for earphones is the electrodynamic (or orthodynamic) principle. This uses a diaphragm which has a coil built in, using printed circuit board techniques. The coil can be a simple spiral design, or a more complicated shape (for better linearity), and the advantage of the method is that the driving force is more evenly distributed over the surface of the diaphragm. Headphones based on this principle have been very successful and of excellent quality.

The ribbon principle is also used to provide loudspeaker action. The moving element of a ribbon loudspeaker is necessarily small, and for that reason the unit is more often confined to high-frequency use (a tweeter) rather than for full-range reproduction. Wide-range multi-ribbon units are also feasible, but in a very different size (and price) category. The commercially available types use three units, of which the bass unit is very large, and which requires its own amplifier to supply about 100–1000 W driving power. The highest power levels are used by sub-woofers, with a restricted frequency range around 20–50 Hz.

The piezoelectric principle has also been used for earphones in the form of piezoelectric (more correctly, pyroelectric, since the electrical parameters are temperature sensitive) plastics sheets which can be formed into very flexible diaphragms. The moving mass is very small and sensitivity is high with no need for a power supply, so that the device is passive. The linearity is not particularly good.

The Quad wide-range electrostatic loudspeaker is one of the very few examples of full-range loudspeakers using the electrostatic principle, although the system has also been used in earphones which, despite the need to provide a high polarizing voltage for the plates, have been very popular and of outstanding audio quality. The advantage that makes the electrostatic loudspeaker principle so attractive is that the driving effort is not applied at a point in the centre of a cone or diaphragm, but to the whole of a surface that can be large in area, satisfying the requirement of moving a large area of air. Electrostatic earphones using electrets permit high-quality listening without the need for a high voltage to be supplied.

Ultrasonic waves

Some transducers that are used for sending or receiving ultrasonic signals through solids or liquids can operate in either direction if required, but for ultrasonic signals sent through the air (or other gases), the transducers are used with diaphragms and in enclosures that can make the application more specialized so that a transmitter or a receiver unit has to be used for its specific purpose. The important ultrasonic transducers are all piezoelectric or magneto-strictive, because these types of transducers make use of vibration in the bulk of the material, as distinct from vibrating a motor unit which then has to be coupled to another material.

Magnetostriction is the change of dimensions of a magnetic material as it is magnetized and demagnetized. Several types of nickel alloys are strongly magnetostrictive, and have been used in transducers for the lower ultrasonic frequencies, in the range 3–100 kHz. A magnetostrictive transducer consists of a magneto-strictive metal core on which is wound a coil. The electrical waveform is applied to the coil, whose inductance is usually fairly high, so restricting the use of the system to the lower ultrasonic fre-quencies. For a large-enough driving current, the core magneto-striction will cause vibration, and this will be considerably intensified if the size of the core and the operating frequency are matched so as to achieve mechanical resonance. The main use of magnetostrictive transducers has been in ultrasonic cleaning baths, as used by watchmakers and in the electronics industry.

The piezoelectric transducers have a much larger range of application, although the maximum possible power output cannot approach that of a magnetostrictive unit (which can be constructed to any size; there is a limit to the practical size of a crystal and to its internal dissipation). The transducer crystals are barium titanate or quartz, and these are cut so as to produce the maximum

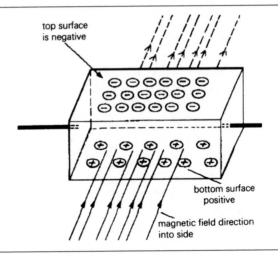

Figure 8.7   The Hall effect. The size of the voltage that is produced by the effect
of a magnetic field is large only when semiconductor materials are used.

vibration output or sensitivity in a given direction. The crystals are
metallized on opposite faces to provide the electrical contacts, and
can then be used either as transmitters or as receivers of ultrasonic
waves. The impedance levels are high, and the signal levels will be
millivolts when used as a receiver, a few volts when used as a transmitter.

Magnetic transducers

Changes in magnetic fields can be detected using a coil, because a
voltage will be induced in a coil when the magnetic flux through
the coil changes. Modern methods of measuring magnetic fields,
however, all rely on the Hall effect. The Hall effect is an example
of the action of a magnetic effect on moving charged particles, such
as electrons or holes, and it was the way in which hole movement
in metals and semiconductors was first proved. The principle is a
comparatively simple one, but for most materials detecting the
effect requires very precise measurements.

   The principle is illustrated in Figure 8.7. If we imagine a slab of
material carrying current from left to right, this current, if carried

entirely by electrons would consist of a flow of electrons from right to left. Now for a current and a field in the directions shown, the force on the conductor will be upwards, and this force is exerted on the particles that carry the current, the electrons. There should therefore be more electrons on the top surface than on the bottom surface, causing a voltage difference, the Hall voltage, between the top and bottom of the slab. Since the electrons are negatively charged, the top of the slab is negative and the bottom positive. If the main carriers are holes, the voltage direction is reversed.

The Hall voltage is very small in good conductors, because the particles move so rapidly that there is not enough time to deflect a substantial number in this way unless a very large magnetic field is used. In semiconductor materials, however, the particles move slowly, and the Hall voltages can be comparatively substantial, enough to produce a measurable voltage of a few µV for only small magnetic fields such as the horizontal component of the earth's field. Small slabs of semiconductor are used for the measurement of magnetic fields in Hall-effect fluxmeters and in electronic compasses. A constant current is passed through the slab, and the voltage between the faces is set to zero in the absence of a magnetic field. With a field present, the voltage is proportional to the size of the field.

Photocells

A photocell is a light-to-electrical transducer, and there are many different types available. Light is an electromagnetic radiation of the same kind as radio waves, but with a very much shorter wavelength and hence a much higher frequency. Light radiation carries energy, and the amount of energy carried depends on the square of the amplitude of the wave. In addition, the unit energy depends on the frequency of the wave. The sensitivity of photocells can be quoted in either of two ways, either as the electrical output at a given illumination, using illumination figures in units of lux, often 50 lux and 1000 lux, or as a figure of power falling on the cell per square centimetre of sensitive area, a quantity known as irradiance. The lux figures for illumination are those obtained by using photometers, and a figure of 50 lux corresponds to a 'normal' domestic lighting level good enough for reading a newspaper. A value of 1000 lux is the level of illumination required for close inspection work and the reading of fine print; on this scale, direct sunlight registers at about 100 000 lux. The use of milliwatts per square centimetre looks more comprehensible to anyone brought up with electronics, but there is no simple direct conversion between power per square centimetre and lux unless other quantities such as spectral composition (colour balance) of light are maintained constant. For the range of wavelengths used in photocells, however, you will often see the approximate figure of $1 \text{ mW/cm}^2 = 200$ lux used.

Another important point relating to the use of photocells is that they are not uniformly sensitive at all visible colours. For many types of sensors, the peak sensitivity may be at either the red or the violet end of the visible spectrum, and some sensors will have their peak response for invisible radiation either in the infrared or the ultraviolet. A few devices, notably some silicon photodiodes, have their peak sensitivity for the same colour as the peak sensitivity of the human eye. The main classes of photocells are photoresistors, photovoltaic materials, and photoemitters.

Photocell types

The photoemissive cell was the dominant type of photosensor for many years. This is a vacuum device, and since it requires a power supply for operation is not strictly a passive device. In addition, the use of photoemissive cells is now rather specialized. The photoresistor also needs a DC supply, and consists of a material which has high resistance in the absence of light, and a much lower resistance when the material is illuminated.

The effect of light on a photodiode is to generate electron-hole pairs in a reverse-biased junction, and the result is that current can flow when light strikes the junction, equivalent to a decrease in the reverse resistance.

PHOTORESISTORS

The most common form of photoresistive cell is the cadmium sulphide cell, named after the material used as a photoconductor.

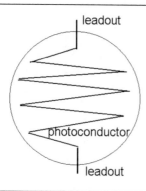

leadout

photoconductor

leadout

Figure 8.8 A cadmium sulphide cell showing the shape of the track of photoconductive material.

This is often referred to as an **LDR** (light-dependent resistor). The cadmium sulphide is deposited as a thread pattern on an insulator, and since the length of this pattern affects the sensitivity, the shape is usually a zigzag line (Figure 8.8). The cell is then encapsulated in a transparent resin or encased in glass to protect the cadmium sulphide from contamination from the atmosphere. The cell is very rugged and can withstand a considerable range of temperatures, either in storage or during operation. The voltage range can also be considerable, particularly when a long track length of cadmium sulphide has been used, and this type of cell is one of the few devices that can be used with an AC supply. One less-welcome feature is that there is a considerable time lag (of the order of several hundred milliseconds) on turning off (increasing resistance when the light is extinguished or dimmed). The switch-on time is less, typically one-fifth of the turn-off time.

PHOTODIODE AND PHOTOTRANSISTOR

A photodiode can be regarded as a high-impedance non-ohmic photosensitive device whose current is almost independent of applied voltage. The incident light falls on a reverse-biased semiconductor junction, and the separation of electrons from holes will allow the junction to conduct despite the reverse bias. Photodiodes

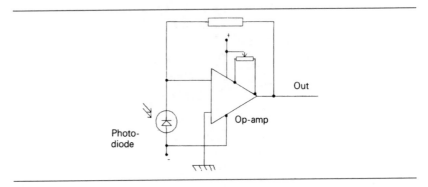

Figure 8.9 The output of the photodiode is normally very small, and amplification is almost always needed.

are constructed like any other diodes, using silicon, but without the opaque coating that is normally used on signal and rectifier diodes. The junction area may be quite large, so that photodiode may have more capacitance between electrodes than a conventional signal diode. This can be compensated by using a feedback capacitor in the circuit, illustrated in Figure 8.9, which shows a typical circuit for using a photodiode along with an operational amplifier for a voltage output. The feedback resistor R will determine the output voltage, which will be RI, where I is the diode current.

A phototransistor is a form of transistor in which the base–collector junction is not covered and can be affected by incident light – it is virtually a photodiode with a base–emitter junction added. The base–collector junction acts as a photodiode, and the current in this junction is then amplified by the normal transistor action so as to provide a much larger collector current, typically one thousand times greater than the output current of a photodiode. The bandwidth is, however, lower. A considerable range of current gain and bandwidth can be obtained. The phototransistor is therefore not a passive component and will not be considered further here.

PHOTOVOLTAIC CELLS

Photovoltaic devices are the only truly passive type of photocells. The first form of photovoltaic device was the selenium cell as used in early types of photographic exposure meters. The principle is that the voltage across the cell depends logarithmically on the illumination, and since for the selenium cell the voltage was of an appreciable size (of the order of 1 V or more in bright illumination) an exposure meter using this type of cell needed no amplification, and could use a meter of reasonably rugged construction. Modern photovoltaic devices are constructed from silicon, and the construction method is as for a photodiode. A silicon photovoltaic device is a silicon photodiode with a large area junction and used without bias. It is connected into a large load resistance, and the typical voltage output is of the order of 0.5 V for bright artificial illumination of 1000 lux.

A more recent application for silicon photovoltaic cells is in electricity generation, using cells in series with a typical output of 0.25 V each, generating enough current to be used to recharge batteries. Series–parallel units can be used to supply DV at a higher level that can be inverted into AC 220 V in areas where no other source of supply is possible. This system can be used for powering remote sensors such as weather stations, and is widely used in satellite power supplies (where the sunlight can be intense and constant).

Electrical to light transducers

LEDs are diodes in which conduction causes the emission of light in the visible range. The majority of LEDs in use are of gallium phosphide or gallium arsenide phosphide construction and are electrically diodes with a high forward voltage of about 2 V, depending on type, and a very small peak reverse voltage of about 3 V. In use, then, great care is needed to ensure correct polarity of connection so as to avoid burning out the LED. Limiting resistors are usually wired in series to prevent burn-out due to excessive current.

The intensity of illumination from the LED depends fairly linearly on the forward current, and this can range from 2 to 30 mA, depending on the physical size of the unit and the brightness required. A few types can operate at 100 mA. The colour of the LED is determined by the material, and the two predominant colours are red and green, although blue can now be obtained. The use of red and green sources close to each other can be used to give yellow light, in accordance with the rules on mixing of light colours, so that the colours red, green, yellow and white can easily be obtained from a single LED package. The use of twin diodes in a package can allow switchable red, green and yellow lights to be obtained, and when LEDs are supplied as part of a package with IC digital units, flashing LED action can be obtained.

Single LEDs are obtainable in a considerable variety of physical forms, of which the standard dot and bar types predominate. The bar type of LED, using semiconductor material in bar form rather than a set of dots, can usually be obtained in intensity matched form, in which the intensity at a specified current rating is matched closely enough to allow the units to be stacked to be used as a column display. In a display of this type, a diode whose intensity is higher or lower than the others is particularly obvious, hence the need for intensity matching.

Opto-couplers and opto-isolators make use of an LED and a phototransistor in one package so that the light output of the LED is the light input for the phototransistor. Since the signal coupling is by way of a light path, which can be of whatever length the design requires, there can be a very high degree of isolation between electrical input and electrical output. The opto-isolator is a more critical component, using the isolation for electrical safety purposes; the opto-coupler, used to pass signals between very widely differing DC levels, requires less isolation.

Thermocouples

A thermocouple is a temperature (or, more correctly, a temperature difference) transducer with an electrical output. The principle is that two dissimilar metals always have a contact potential between them, and this contact potential changes as the temperature changes. The contact potential is not measurable for a single connection (or junction), but when two junctions are in a circuit (Figure 8.10) with the junctions at different temperatures then a voltage of a few microvolts per degree can be detected. This voltage will be zero if the junctions are at the same temperature,

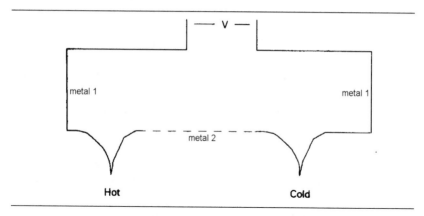

Figure 8.10 A thermocouple junction exists where two differing metals meet, and when two junctions exist in a circuit this produces a voltage of a few millivolts when there is a temperature difference between the two junctions.

and a plot of voltage against temperature difference is approximately linear.

- There are always several junctions in a circuit, because a junction exists wherever two different conducting materials meet, and if copper wire is used for the circuit there will be a junction where one of the thermocouple metals joins to the copper. The essence of a thermocouple, however, is that these other junctions are all at a constant temperature and make no net contribution to the output voltage.

Some combinations of metals will give a parabolic characteristic, and are therefore not used for thermocouples. Most materials will provide a linear 0°C to 100°C range and combinations such as platinum–rhodium provide much larger linear ranges. The output from a thermocouple is small, at most a few millivolts for a 100°C temperature difference, and more typically a fraction of a millivolt. If the output of the thermocouple is required to drive anything more than a meter movement, then DC amplification will be needed, using an operational amplifier or chopper amplifier.

RESISTANCE THERMOMETERS

Resistance thermometer transducers, also known as RTDs (resistance temperature detectors) depend on the change of resistivity of any material when the temperature changes, and this change in the resistivity causes a change of resistance. The resistance change is not linear, particularly over a large range of temperatures, but the linearity can be improved by using an active circuit with a small amount of positive feedback. Resistance thermometers are often used in applications that call for the measurement of very high temperatures.

For comparatively small temperature ranges, up to $400°C$ or so, the resistance change of nickel or of nickel alloys can be used, and for higher temperature ranges, platinum and its alloys are more suitable because of their much greater resistance to oxidation. For measurement purposes, the resistance sensor can be connected to a measuring bridge, along with a dummy sensor whose temperature is kept constant. The platinum-resistance thermometer is a standard type, calibrated from the ultimate but inconvenient temperature reference of a gas expansion thermometer. Platinum sensors can now be obtained in thin-film or thick-film forms of various shapes and sizes, at much lower cost than the older platinum-wire types.

Thermistors

Thermistors are a form of temperature-sensitive resistor formed using mixtures of oxides of exotic metals. The constructional methods are similar to those used for carbon composition resistors. Some of these mixtures have positive temperature coefficients, and in most cases it would be meaningless to quote a value for temperature coefficient, positive or negative, because the value is not a constant. The thermistors with a positive temperature coefficient are very non-linear, but the more common negative temperature coefficient types follow a roughly logarithmic law with no violent changes in resistance.

Given that the resistance of a thermistor is known at one temperature θ_2, it can be calculated for another temperature θ_1 by using the formula which is illustrated in Table 8.1. The use of θ rather than T for temperature in this formula is a reminder that the temperatures must be in units of Kelvins (absolute temperatures). The Kelvin or absolute temperature is obtained by adding 273 to the

Table 8.1 Thermistor formulae that relate resistance to temperature

The temperature coefficient of a thermistor is not a constant, but itself varies as temperature changes. A more useful quantity is the thermistor constant B which can be used to find the resistance at any temperature in the working range provided another pair of resistance and temperature values are known.

The equation is

$$R_2 = R_1 \cdot e\left(\frac{B}{\theta_1} - \frac{B}{\theta_2}\right)$$

θ and B values are in Kelvin (K) units of temperature.

For example, if the thermistor constant B is known to be 3200 K, and the resistance at 30°C is 2 k, then the resistance at 45°C can be calculated as follows:

The temperatures are 293 K and 318 K, so that the quantity in brackets is 0.8586. Using the EXP function of a calculator, $R_2 = 2 \times 2.359 = 4.719$ about 4k7.

Celsius temperature. If you need to work to two places of decimals of temperature, use the figure 273.16.

Thermistors can be obtained in a variety of physical forms, as beads, miniature beads, plates, rods, and also encapsulated in metal containers.

NTC thermistors are used for temperature control applications such as low-temperature oven controllers, deep-freezer thermostats, room temperature sensors and process controllers. Temperature limits range from 150°C to 200°C, with a few types able to withstand 600°C. The range of temperature that a thermistor can handle depends on the associated circuit, because the range of resistance will be very large compared to the range of temperature.

In any of these applications, NTC thermistors have considerable advantages as compared to the old bimetal thermostat, notably the absence of any hysteresis effects (switching on at a different temperature than that for switching off). NTC thermistors can also be obtained in evacuated envelopes for use in such purposes as oscillator limiters and controllers for voltage-controlled amplifiers.

Thermistor circuits in general require the use of preset potentiometers in order to make working adjustments, but circuit costs can be reduced by employing curve matched thermistors whose resistance values are guaranteed to within close limits at each of a large range of temperatures. Thermistors of all types will also have quoted values of dissipation constant and time constant. The dissipation constant is the amount of power (in milliwatts) that is required to raise the temperature of the thermistor by 1°C above the ambient temperature. For the evacuated bulb type, the dissipation constant is very small, of the order of 12 µW/°C, so that the resistance of this type of thermistor is substantially altered by only very small amounts of signal current. For temperature-sensing thermistors, values of dissipation constant in the range 70–500 µW/°C are typical.

The time constant for a thermistor is defined as the time needed for the resistance to alter by 63% of the difference between an initial value and a final value caused by a change of temperature. Time constant is measured with negligible current flowing, because otherwise the figure would be altered because of part of the heating being internal rather than external. The figure of 63% may seem odd, but it corresponds to the definition of time constant for other networks such as a capacitor and a resistor, and by making the definition in this way, the value of time constant is genuinely constant over a large range of temperature changes. Time constants of 5–11 seconds are typical of the physically small thermistors (miniature beads and the evacuated bulb types), with much larger values for others, 18–25 seconds for the larger beads, and as high as 180 seconds for thermistors that have been assembled into temperature-sensing probes.

Negative temperature coefficient (NTC) thermistors can be constructed from semiconducting materials with values of temperature coefficients that are usually much larger than the (positive) temperature coefficients of resistors. The phrase NTC resistor is used for devices with fairly small negative values of temperature coefficient, and the term thermistor is reserved for the types that have large negative values of temperature coefficient. Most of the thermistors that are incorporated into temperature-sensing circuits are of the NTC type.

PTC THERMISTORS

Positive temperature coefficient (PTC) thermistors are a more recent development, used mainly for protection circuits for sensing temperature or current. Unlike the NTC types, these PTC thermis-

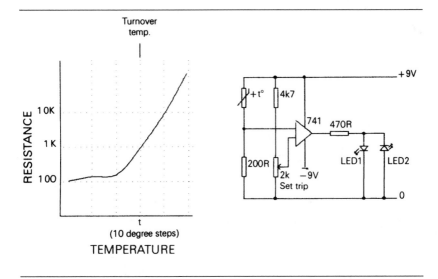

Figure 8.11 Characteristic for a PTC thermistor used for sensing excess temperature.

tors have a current–voltage characteristic that exhibits a change in direction, and two basic types are used, both depending on compounds of barium, lead and strontium titanates (ceramic materials). The overtemperature protection type of PTC device has a switch-over point at a reference temperature (or trip temperature, T_r). At temperatures lower than the trip temperature the resistance of the PTC device is fairly constant, but around the trip temperature the PTC characteristic takes over and their resistance rises very sharply as the temperature rises. A typical graph of resistance plotted against temperature, along with a trip-sensing circuit, is shown in Figure 8.11. The sudden change in resistance can be used to operate an indicator or to switch other circuits for purposes such as motor protection or for preventing overheating of transformers.

The other type of PTC thermistor device is used for overcurrent protection circuits, and its resistance/temperature graph is illustrated in Figure 8.12. This characteristic follows an S-shaped curve which has two turn-over points, one at a point of minimum resistance R_{min} and the other at the maximum resistance point R_{max}. Between $0°C$ and R_{min}, the temperature coefficient is negative, and the coefficient is also negative at the temperatures in the region higher than R_{max}. Between R_{min} and R_{max} the tempera-

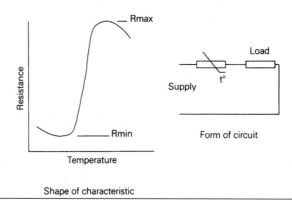

Shape of characteristic

Figure 8.12   The type of characteristic used for current protection devices operated by a PTC thermistor.

ture coefficient is large and positive. In this PTC region, the change of resistance can be as large as 100% for each degree Celsius rise in temperature.

With one of these devices wired in series with a load, the load is protected against excessive current. At the working current, the PTC device is in its low-resistance state, allowing most of the applied voltage to be across the load. When the current is increased, the thermistor will suddenly switch to its PTC mode, assisted by self-heating as more of the applied voltage will now be across the thermistor, until the current flowing in the whole circuit becomes very small. The circuit can be arranged so that it is self-resetting, returning to normal when the thermistor cools, or requires the circuit to be reset by switching off the current and allowing the thermistor to cool.

• Note that if an op-amp with sufficient gain is used to amplify the voltage across a thermistor, any type of thermistor can be used to sense excessive temperature. The advantage of using a PTC thermistor is that many types of sensing applications can use circuits that require no op-amp.

Mechanical quantities

A resistive strain gauge consists of a conducting material in the form of a thin wire or strip which is attached firmly to the material in

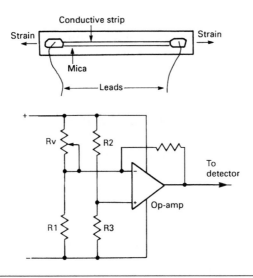

Figure 8.13 The strain gauge principle. The strain gauge element can be made from a metal strip, but silicon is much more common because of its much higher sensitivity.

which strain is to be detected. This material might be the wall of a building, a turbine blade, part of a bridge, anything in which excessive stress could signal impending trouble. The fastening of the resistive material is usually by means of epoxy resins such as Araldite, but ceramic pastes are used for high-temperature applications. Electrically, the strain gauge strip is connected as part of a resistance bridge circuit (Figure 8.13). The effects of temperature can be minimized by using another identical unstrained strain gauge in the bridge as a comparison.

The use of a semiconductor strip in place of a metal wire makes measurement much easier, because the resistance of such a strip can be considerably greater, and so the changes in resistance can be correspondingly greater. Except for applications in which the temperature of the element is high (gas-turbine blades, for example), the semiconductor type of strain gauge is preferred.

Piezoelectric strain gauges are useful where the strain is of short duration, or rapidly changing in value. The voltage can be very large, of the order of several kilovolts for a heavily-strained crystal, so that the gauge can be sensitive, but the output impedance is very high and capacitive. The output is not DC, therefore, so that this type of gauge is not useful for detecting slow changes, and its main application is for acceleration sensing.

Surface acoustic wave devices

Quartz crystals, cut into thin plates and with electrodes plated on to opposite flat faces, can be used as resonant circuits with Q values ranging from 20 000 to 1 000 000 or more. The equivalent circuit of a crystal is shown in Figure 8.14. The crystal by itself acts as a series resonant circuit with a very large inductance, small capacitance and fairly low resistance (a few thousand ohms). The stray capacitance across the crystal will also permit parallel resonance to occur at a frequency that is slightly higher than that of the series resonance. Figure 8.15 shows how the reactance and the resistance of a crystal vary as the frequency is changed – the reactance is zero at each resonant frequency and the resistance is maximum at the parallel resonant frequency. Usually the parallel and the series resonant frequencies are specified when the crystal is manufactured. In these applications, the quartz crystal is acting as an electrical-to-acoustic transducer, and a development of this principle uses transducer crystals that have a reversible action, so that the acoustic-to-electrical conversion is used also.

A surface acoustic wave (SAW) device is an arrangement that uses two transducers back to back with electrical input and electrical output but with an acoustic wave (which is usually ultrasonic)

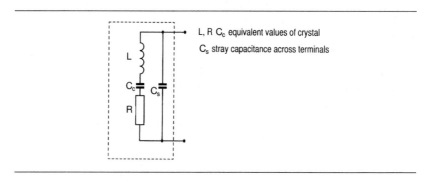

Figure 8.14 Equivalent circuit of a quartz crystal.

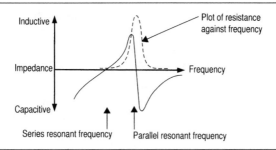

Figure 8.15 Variation of reactance and resistance of a quartz crystal near its resonant frequencies.

between. A signal input to one transducer causes the wave to be set up, and the wave reaching the second transducer causes an electrical output at the same frequency, but delayed compared to the original signal. The device can be used purely as a signal delay, and the best-known example is the delay line originally used for decoding the PAL colour TV signal. This is made from glass cast into a V shape (Figure 8.16), with one transducer converting the video signal into an ultrasonic wave which reflects and hits the other transducer some time later. The precise time can be adjusted by trimming the reflecting end of the glass.

Delay lines of this type can give delays of milliseconds as compared to the microseconds that can be achieved with electrical lines or with GC networks. The output signal is very considerably attenuated as compared to the input signal, so that amplification is

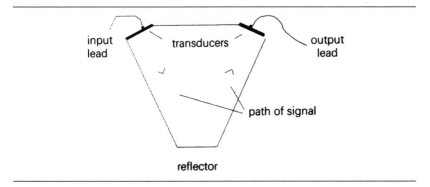

Figure 8.16 The form of a glass delay line as used for early PAL colour TV receivers. The delay can be adjusted by grinding the apex.

needed both before and following the delay line. This type of analogue delay device has now been superseded by digital delays which can be adjusted more easily.

The other main SAW application is filtering. These are used wherever a steep cut-off form of filter is needed without the problems of ringing that would appear if a conventional LC filter were used; that is, if a conventional LC filter could be designed to achieve such steep cut-off. Typical applications are in the PAL TV system to separate subcarriers, and in CD players to avoid 'aliasing' due to harmonics.

If the transducers are formed on a crystal which has a mechanical resonance, the output will be strongly frequency dependent, and the equivalent Q of the arrangement can be very high, almost in the quartz crystal regions of 30 000 or so. An added advantage is that by careful mechanical design, band-pass filters can be constructed which have a very steep transition between pass-band and stop-band, very much better than can ever be achieved with any LC combination, active or passive.

Chapter 9

Hardware

The hardware of electronics, meaning the chassis, covers, cabinets, switches and other 'external' features, is almost as important as the circuitry within, because the hardware is all that the casual user can use to judge the system. British manufacturers of hi-fi systems (about all that survives of the UK consumer electronics industry) are by now well aware of this, although it seems that we have the strange distinction of making as much valve-equipped hi-fi as transistorized versions. For consumer applications, the look of hardware is important, but for military contracts, adhering to specifications is what counts. On all scores hardware cannot be neglected, although it very often is. The task is made much easier when there is a company policy of using some particular form of hardware such as standardized board and cabinet sizes. The most difficult decision is on how to package some piece of equipment that is for the moment a one-off product, particularly if there is any chance that other items will follow.

Terminals and connectors

The prospective user of a piece of electronics equipment first makes contact with the design when he/she tries to connect it to other equipment and to whatever power supply is used. Mains-operated equipment for domestic or office use will have a connected and well-tethered mains cable of the correct rating, preferably with a correctly fused 3-pin plug if it is intended for the UK market. The option is to use a BS/IEC approved 3-pin plug on the chassis, with a lead that has a matching socket at one end and a suitable mains plug at the other. Mains cables can be obtained as a standard item with the BS/IEC fitting at one end and various UK or other plugs at the other. If additional equipment has to be driven, BS/IEC sockets can be used to allow power to be taken from the main unit rather than from a set of additional mains cables. Only low-power and double-insulated equipment should use the 2-pin form of cassette recorder leads because these provide small contact area and are not mechanically well secured.

Industrial equipment for UK use will use one of the approved industrial connectors, almost certainly to the BS 4343 specification for equipment to be used in the UK. The BS scheme specifies two digits, but the third which is illustrated can be added to show protection against mechanical impact.

Mains connections are about as standardized as anything in the use of terminals ever attains. When we look at what is available for low-voltage connectors and for signal connectors, there is a bewildering variety of styles available. The primary aim in choosing connections should be to achieve some co-ordination with the equipment that will be used along with yours, and this means that you need to have, from the start of the design stage, a good idea of what amounts to de facto standards. If no such standards exist, try to resist making new ones because there are far too many already.

One example is the low-voltage supply connector which is used for cassette recorders, portable stereo players and for some calculators. Even in this very restricted range of applications there are two sizes of plug/socket, the 2.1 mm and the 2.5 mm, with some manufacturers using the centre pin as ground and others using the centre pin as the supply voltage pin (usually 6 V), so that there can be four variations on this design alone. Power supplies for such equipment usually cope by using a lead that is terminated in a 4-way jack plug and which at the supply end is fitted with a reversible plug, often with no polarity markings.

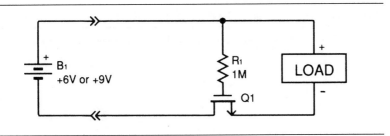

Figure 9.1 A circuit due to Bob Pease which will protect a load against accidental voltage reversal.

Getting the correct polarity is very often more a matter of luck than good instructions, so that if this type of connector is used there should be a clear indication of polarity and also some protection against use of the wrong polarity (a diode in series) if this can be tolerated. In some cases the use of a diode imposes an excessive voltage drop and protection is very difficult to achieve at reasonable cost. A more elaborate method, with a much lower voltage drop, is illustrated in Figure 9.1 (due to Bob Pease). The voltage drop across the FET is much less than the 0.5 V (minimum) across a diode.

The main confusion, however, exists among signal connectors, and the only possible advice here is to try to stay with industry standards for comparable equipment. For educational equipment, for example, the 4 mm plug and socket is almost universal, and for connecting UHP signals to a domestic TV receiver the standard coaxial plug and socket type should be used. Connections to TV receivers that are being used as monitors often use the SCART form of Euroconnector, see later, but for other connections, particularly for computer monitors, standards can vary widely. Fortunately, the almost universal adoption of the IBM PC standards everywhere except in education (where they are most needed) ensures reasonable uniformity.

The largest range of connectors is found in the TV and video ranges, with audio coming a close second. RF connectors are used for radio transmitters, including CB and car telephone uses, and for a variety of VHF and UHF work, and although virtually all coaxial in design, offer a wide range of fittings, whether bayonet or screw retained. The wide range of connectors reflects the wide range of VHF and UHF cables, so that you cannot necessarily use

Table 9.1   A selection of the RG standards for RF cables

The most commonly-used types of RF cables can be classed into six groups as follows:

Gp.	Example	Impedance	pF/metre	Attenuation/frequency
A	RG58C/U	50	100	2/10 3.1/200 7.6/1000
B	RG174A/U	50	100	1.1/10 4.2/200 6.0/400
C	RG223/U	50	96	0.39/10 1.58/100 5.41/1000
D	UR M70	75	67	1.5/100 5.2/1000
E	RG179B/U	75	64	1.9/10 3.2/100 8.2/1000
F	RG59B/U	75	68	1.3/100 1.9/200 4.6/1000

The UR series of cables are to UK Uniradio standards, the RG set are to US standards. Capacitance is quoted in picofarads per metre length, and the attenuation is shown in db per 10m length of cable in MHz, at selected frequencies.
For example, 5.2/1000 means 5.2 db attenuation at 1000 mHz for a 10m length of cable.

any type of connector with any type of cable. Generally, connectors are made to work with a limited range of cables, although in some cases the range can be extended by using adapters.

The range of RF connectors is intended to match the range of RF cables, of which Table 9.1 shows a summary only, in which some of the better-known cable types are arranged into groups. RF cables are identified by Uniradio numbers or by RG numbers. Both systems of classification are in widespread use in the UK, but the RG set, of US origin, is better known worldwide. The main measurable features of an RF cable are the characteristic impedance to which the cable must be matched in order to avoid reflections, the capacitance per metre, and the attenuation in dB per 10 m at various RF frequencies. All of the cable groups illustrated are coaxial and the characteristic impedance is virtually always either 50 Ω or 75 Ω, but the attenuation characteristics differ considerably from one cable type to another. Most of the RF cables are rated to withstand high voltages between inner and outer conductors, often exceeding 20 kV.

- Characteristic impedance values of 53 Ω and 93 Ω are used in the USA, and another standardized impedance, used for microstrip, is 100 Ω.

The various connectors are likely to be used for other than RF cable connectors, of course, and there is a wide range of other coaxial cables which will fit one connector type or another and to which the connectors will match well. These applications include audio, video and digital network signal applications.

The BNC range of connectors covers both 50 Ω and 75 Ω types which are manufactured for an assortment of cable sizes. All feature a bayonet locking system, and a maximum diameter of about 15 mm, and both solder and crimp fittings are available. The standard range of BNC connectors can be used in the frequency range up to 4 GHz (absolute maximum 10 GHz) and with signal voltage levels up to 500 V peak. Terminations and attenuators in the same series are also obtainable, and there is also a miniature BNC type, 10 mm diameter, with the same RF ratings, and a screw-retained version, the TNC couplers. The connectors of this family offer a substantially constant impedance when used with the recommended cables.

The SMA series of connectors are to BS 9210 N0006 and MIL-C-39012, and are screw retained. This provides more rigidity and improves performance under conditions such as vibration or impact. The voltage rating is up to 450 V peak, and frequencies up to 12.4 GHz on flexible cable, and up to 18 GHz on semi-rigid cable, are usable. The VSWR (voltage standing wave ratio, ideally equal to 1) which measures reflection in the coupling is low at the lower frequencies but changes with frequency, usually showing a peak at some frequency. A typical quoted formula for semi-rigid cable coupling is $1.07 + 0.008f$ with f in GHz, so that for a 10 GHz signal the VSWR would be $1.07 + 0.008 \times 10 = 1.15$. The body material is stainless steel, gold plated, with brass or beryllium–copper contacts and a PTFE insulator. The typical operating temperature range is $-55°C$ to $+155°C$.

The SMB (sub-miniature bayonet) range is to BS 9210 and MIL-C-93012B specifications and has a 6 mm typical diameter, rated for 500 V signal peak in the frequency range up to 4 GHz. The VSWR is typically around 1.4 for a straight connector and 1.7 for an elbowed type, and both solder and crimp fittings are available. SMC (subminiature screw) connectors are also available to the same BS and MIL specifications.

The older UHF plug series are also to MIL specifications, but with the limited frequency range up to about 500 MHz. These are larger connectors, typically 19 mm diameter for a plug, with screw clamping, and they are particularly well suited to the larger cable sizes. They are often used for video signal coupling.

There are adapters available for every possible combination of RF connector, so that total incompatibility of leads should never arise. This is not a perfect solution, however, because the use of an adapter invariably increases the VSWR figure, and it is always better to try to ensure that the correct matching plug/socket is used in the first place.

Video connectors

Video connections can make use of the VHF type of RF connectors, or more specialized types. These fall into two classes, the professional video connectors intended for use with TV studio equipment, and the domestic type of connector as used on video recorders to enable dubbing from one recorder to another or from a recorder to a monitor so that the replayed picture can be of better quality than is obtainable using the usual RF modulator connection to the aerial socket of TV receivers. Many video recorders nowadays use nothing more elaborate than an audio style (RCA) phono connector for their video as well as their audio output, and a SCART socket for connection to the TV receiver, with possibly another SCART for connection to satellite or other digital decoder.

For studio use, video connectors for a camera may have to carry a complete set of signals, including separate synchronizing signals, audio telephone signals for a camera operator, and power cables as well as the usual video-out and audio-out signals. Multiway rectangular connectors can be used for such purposes, with 8-way for small installations and 20-way connections for connections to editing consoles and similar equipment. These connectors feature very low contact resistance, typically 5 mΩ. For smaller equipment, circular cross-section connectors of about 17 mm overall diameter are used, carrying ten connectors with a typical contact resistance of 14 mΩ and rated at 350 V AC.

Audio connectors start with the remarkably long-lived jacks which were originally inherited (in the old $\frac{1}{4}$ in size) from telephone equipment. Jacks of this size are still manufactured, both in mono (2-pole) and stereo (3-pole) forms, and with either chassis mounted or line sockets. Their use is now confined to professional audio equipment, mainly in the older range, because there are more modern forms of connectors available which have a large contact area in comparison to their overall size. Smaller versions of the jack connector are still used to a considerable extent, however, particularly in the stereo form. The 3.5 mm size was the original miniature jack, and is still used on some domestic equipment, but the 2.5 mm size has become more common for mono use in particular.

One of the most common forms of connector for domestic audio is still the phono connector, devised by RCA, whose name indicates its US origins. Phono connectors are single channel only, but are well screened and offer low-resistance connections along with sturdy construction. The drawback is the number of fittings needed for a 2-way stereo connection such as would be used on a stereo recorder, and for such purposes DIN plugs are more often used, particularly in European equipment. Many users prefer the phono type of plug on the grounds of lower contact resistance and more secure connections. At one time, DIN connectors were almost unknown in the USA, but they have been introduced in computing applications, notably as the keyboard connector for the older IBM PC machines. The later PS/2 type of connector is similar in size and style, but with a quite different pin arrangement and a different locating system.

The European DIN (Deutsches Industrie Normallschaft, the German standardizing body) connectors use a common shell size of 15 mm diameter for a large range of connections from the loudspeaker 2-pole type to the 8-way variety. Although the shell is common to all, the layouts (Figure 9.2) are not. The original types are the

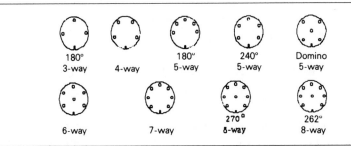

Figure 9.2 The various 15 mm diameter DIN connector pin layouts for audio uses. There is also a 2-pin type (one round pin, one flat pin) for loudspeaker connections.

3-way and the $180°$ 5-way connectors, which had the merit of allowing a 3-way plug to be inserted into a 5-way socket. Later types, however, have used $240°$ pin configurations with 5, 6 and 7 contacts; 4-way and ground types with the pins in square format; and a 5-way 'domino' type with a central pin, along with the 8-way type which is configured like the 7-way $240°$ type with a central pin added. This has detracted from the original simplicity of the scheme, which was intended to make the connections to and from stereo domestic audio equipment easier.

The more crowded layouts of DIN plugs and sockets are notoriously difficult to solder unless they have been mechanically well designed, using splayed connectors on the chassis mounted sockets and to some extent also on the line mounted plugs. Use of the 5-way $240°$ type is recommended for audio equipment other than professional grade equipment, but only where signal strengths are adequate and risk of hum pickup is minimal. The standard system of pin assignment should be adhered to if the connector is being used in its typical application, as input/output for a tape/cassette recorder. Latched connectors can be obtained to avoid the possibility of pulling the connectors apart accidentally. For low-level use, phono plugs are preferable.

For professional audio (or high-quality domestic audio) equipment, the XLR series of connectors provides multiple connections with much superior mechanical quality. These are available as 3-, 4-, or 5-pole types with anchored pins to which the cables are soldered, with set-screws used to retain the shells. The contacts are rated to 15 A for the 3-pole design (lower for the others) and they can be used for a maximum working voltage of 120 V. Contact resistance is low, and the connectors are latched to avoid accidental disconnection. There is a corresponding range of loudspeaker connectors to the same high specifications.

- You may need to specify a grounded shell design if the connectors are to be used with shielded cables.

A variety of other connectors also exists, such as the EPX series of heavy-duty connectors and the MUSA coaxial connectors. These are more specialized and would be used only in their specialized applications, they are not found used as general-purpose connectors.

THE SCART CONNECTOR

This device, Figure 9.3 and Table 9.2, is known variously as the
SCART, Peritelevision or Euroconnector. It provides a standard
21-pin interface designed to allow various peripheral devices to be
connected to a TV receiver. Whilst the physical arrangement of the
connector and its pins and some of the connections have been standardized, there are a number of variants found in practice and
these are shown in the table below. The device allows for connections to be made at standard RGB or baseband levels so that no
remodulation is needed to provide input to the UHF aerial socket.
The use of remodulation tends to degrade the picture quality due
to mixer noise and tuner drift.

COMPUTER CONNECTORS

Computer and other digital signal connections have, at least,

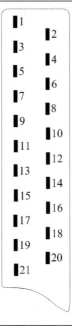

Figure 9.3   The SCART or Peritel socket diagram.

Table 9.2 SCART connector pinout

Pin	Connection	Pin	Connection
1	Right audio channel output	14	Data bus ground
2	Right audio channel input	15	Red video input
3	Left audio channel output	16	Mode switch, 0 – composite
4	Audio screen/ground		video, 1 – fast video blanking
5	Blue video screen/ground	17	Composite video screen/
6	Left channel audio input		ground
7	Blue video input	18	Common earth for pins 8 & 16
8	Source switching, 0 – off air,	19	Composite video output/
	1 – PeriTV		computer sync input
9	Green video screen/ground	20	Composite video input
10	Data bus	21	Cable/connector shell ground
11	Green video input		
12	Data bus		
13	Red video screen/ground		

reached some measure of standardization after a period of chaos. At one time, it was quite common to see edge connectors used for external connections, but this way of breaking circuit boards has now been abandoned, and edge connectors should be confined now to internal connections. Since the advent of the IBM PC style of computer, there are four main types of connector that are peculiar to computing and digital circuits.

The Centronics connector is used mainly for connecting a computer to a printer, and it consists of a 36-contact connector which uses flat contact faces. At one time, both computer and printer would have used identical fittings, but it is now more common for the 36-pin Centronics socket to be used only on the printer. At the computer a 25-pin subminiature D-connection is used, usually with the socket chassis mounted. In a normal connection from computer to printer, only eighteen of the pins are used for signals (including ground). The shape of the body shell makes the connector non-reversible.

The original form of Centronics parallel port was intended for passing signals in one direction only, from a computer to a printer. Several designers made use of the unmodified system for bi-directional signals by using the four signal lines that communicated in the reverse direction along with four data lines so as to get 4-bit bi-directional signalling. This in turn gave rise to a standardized system for allowing the use of the parallel port for 8-bit bi-directional signalling.

This is the standard IEEE Std. 1284-1994 system, and is otherwise known as the EPP (extended parallel port) system. The EPP system is used for modern printers to allow better software control so that, for example, an inkjet printer can signal that it is running low on ink, or a laser printer can signal that it is running low on toner. More significantly, the EPP has been used for industrial applications as an interface between the computer and machines connected to the computer and controlled by it. Printers of recent manufacture also feature a USB connection as an alternative.

The IEEE 1284 standard provides for high-speed signal transfer in both directions between the PC and an external peripheral. The speed of data can be 50 to 100 times faster than was possible using the older Centronics port, but the EPP connection on the PC is still fully compatible with older printers and other peripherals that use the parallel port. You can also use an EPP port along with an older Centronics port on the same PC.

The EPP type of port is standard on modern computers, and should normally be set up by using a (default) option in the CMOS ROM. The system offers five modes of operation, four of which maintain compatibility with older methods:

1. Data in forward direction only (out from computer), used for a normal Centronics printer connection.
2. Bi-directional action using four status lines for data in reverse direction along with four data lines in forward direction. This is also known as *Nibble mode*, and has also been used in cables for connecting two or three computers together in a simple network.
3. Hewlett-Packard Bitronics bi-directional mode, using data lines.
4. ECP (extended capability port) mode for printer and scanner use.
5. Fully bi-directional EPP mode used by some printers and also for computer peripherals such as external CD-ROM, hard drive, etc.

The older bi-directional systems require software to implement each transfer, and this limits the transfer rate to, typically, 50–100 Kbytes per second. Modern PC machines have a port that can be used for ECP and EPP modes, and the I/O controller chip firmware allows for direct control of the port action with a greatly reduced external software overhead. A good comparison is the difference between a DMA (Direct Memory Access) transfer and one made by using the processor to read data and write to memory.

- The IEEE-1284 standard also provides supporting protocols that allow the PC and its peripheral to agree on which mode to use. The standard also defines the cable and connector formats, with electrical signal specifications.

Having EPP/ECP capability on a computer does not guarantee that it will be used when you connect a printer to the port. The printer itself must be capable of operating (usually) in ECP mode, and the operating system must also be capable of using the ECP mode. Although an ECP port can operate in Centronics mode faster than the older type of parallel port, full ECP speed with a printer that can work in this mode requires the ECP port to be set up in the CMOS RAM, and the operating system to use it.

Three types of connectors can be used. One, Type A, is the existing DB25 type, updated to 1284 electrical standards. Type B and C are 36-pin connectors, of which the Type C is the standard that is recommended for new designs. Type C is smaller than older 36-pin types, uses a simple clip as anchor, and permits the use of additional signals, peripheral logic high and host logic high. These additional signals are used to find if the devices at each end of the cable are switched on.

Figure 9.4 shows these connectors; Types A and B are the familiar DB25 and Centronics types that are currently used.

The same 25-pin D-connector can be used for serial connections, but more modern machines use 9-pin sub-miniature D-sockets for this purpose, since no more than 9-pin connections are ever needed. For other connections, such as to keyboards, mice and monitors, DIN style connectors are often used, although the sub-miniature

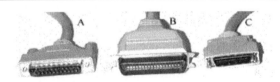

Figure 9.4   Standardized connectors for IEEE-1284 parallel system.

D-type connectors are also common. The D-type connectors are widely available in a range of sizes and with a large range of accessories in the form of casings, adapters and tools, so that their use of all forms of digital signals is strongly recommended.

There are now standard DIN fittings for edge connectors, including the more satisfactory indirect edge connectors that have now superseded the older direct style. The indirect connectors are mounted on the board and soldered to the PCB leads, avoiding making rubbing contacts with the board itself.

THE UNIVERSAL SERIAL BUS (USB)

The essential simplicity of serial connections for linking a computer to a peripheral has spurred designers into looking for something better than the old telecommunications serial ports that belong to the pre-computer age. The requirement was for a type of serial port that could be used for all the normal computer connections, and which could be connected in a 'daisy-chain' type of network with each device connected to another with only one of them needing to be connected directly to the computer.

The answer to this is called the universal serial bus (USB). It has been designed to be *hot-plugged*, meaning that devices can be connected and disconnected with the computer switched on and working. This is possible only if the system is supported by the computer, the peripheral device and the operating system.

USB permits communications between devices that are equipped with suitable interfaces at serial data rates ranging from 1.5 Mbits/s to 1.5 Mbytes/s. This is very much faster than the old-style serial port system. The interconnecting cable has a maximum length of 5 metres and consists of two twisted pair cables, one pair for power and the other for signalling. The distance can be extended to about 30 metres by using a *hub* terminal device as a line repeater.

Terminal devices, such as keyboard, mouse and printer, are added to the basic PC in a daisy-chain fashion and each is identified by using a 7-bit address code. This allows up to 127 devices, in theory at least, to be connected. In practice, not all devices allow daisy-chain connection (picture a mouse with two tails!), and so

Figure 9.5   The two main types of USB sockets used on current models of computers and peripherals.

the computer needs more than one USB connector. Figure 9.5 shows the two main types of USB sockets in current use.

- Other terminal devices can include scanners, fax machines, telephone and ISDN lines, multi-media display and recording systems and even industrial data acquisition devices. A modified, faster, USB-2 system is now available.

FIREWIRE (IEEE 1394-1995)

FireWire is a trademark of Apple Computers Inc. who, during 1988, originally designed and established this type of serial signalling system as the basis for a very fast, low cost and easy to use network for digital audio signals. The system has been developed and standardized by the Institution of Electrical and Electronics Engineers (IEEE) in the USA and is now well suited to be used as an interface for fast computer systems. Since 1995 the network has become an established IEEE standard which is supported by a worldwide trade organization of more than 90 manufacturers and constructors.

FireWire allows any device equipped with a suitable interface to be simply coupled to the computer to form a communication system. This flexibility now allows, for example, the home PC, the television receiver and telephone systems to be connected. Because FireWire allows hot-plugging (you can add or remove connections with the computer switched on) it makes interconnections very quick and simple for the non-technical user.

Any devices that are fitted with the appropriate interface can be coupled together through any one of a number of ports using a simple cable without any consideration for the order in which the devices appear on the network. The services currently available to this network include home video editing, security monitoring, photo-CD handling, image enhancement, video and teleconferencing, plus professional broadcast and industrial applications. Note that these are the applications that need faster transfer of bytes than can be achieved using USB.

FireWire terminal devices may be fitted with single input and multiple output port interfaces which can be coupled together through a special cable unit. Any new device can be added to the network by simply plugging into a spare port anywhere on the network. The devices may be coupled in a mix of clusters or stars or the daisy-chain format, the only restrictions being that there should be no more than sixteen hops between any two nodes and without any loops being formed. The maximum length with no repeaters is about 10 metres, but 4.5 metre lengths are more common.

The most important feature of FireWire is its simplicity, at least as far as the domestic user is concerned. Any new device may be plugged into a spare port without switching the power off and the system then dynamically reconfigures itself to suit the new situation without the need to reset any switches or jumpers.

Control knobs and switches

There is as great a variety in control knobs and switches as in terminals and connectors. Control knobs for rotary potentiometers are available in a bewildering range of sizes and styles, mostly using grub-screw (set-screw) fastening, although a few feature push-on fitting. For all but the least costly equipment, a secure fastening is desirable, but the traditional grub-screw is not entirely satisfactory because it can work loose and cause considerable delay when knobs have to be removed for servicing work. A more modern development is collet-fitting, using a split collet over the potentiometer shaft which is tightened down by a nut. The recess for the collet nut is then covered by a cap which can be colour coded or moulded with an arrow pointer. This is a much more satisfactory form of fitting.

A more specialized form of knob allows multi-turn use, so that ten to fifteen turns of the dial will be needed to rotate the potentiometer shaft from one end stop to the other. These multi-turn dials can use digital or analogue readouts and are normally located by a grub-screw with an Allen (hexagon) head.

Switches

Switches are required to make a low-resistance connection in the ON setting, and a very high-resistance insulation in the OFF setting. The resistance of the switch circuit when the switch is on (made) is determined by the switch contacts, the moving metal parts in each part of the circuit which will touch when the switch is on. The amount of the contact resistance depends on the area of contact, the contact material, the amount of force that presses the contacts together, and also in the way that this force has been applied.

If the contacts are scraped against each other in a wiping action as they are forced together, then the contact resistance can often be much lower than can be achieved when the same force is used simply to push the contacts straight together. In general, large contact areas are used only for high-current operation and the contact areas for low-current switches as used for electronics circuits will be small. The actual area of electrical connection will not be the same as the physical area of the contacts, because it is generally not possible to construct contacts that are precisely flat or with surfaces that are perfectly parallel when the contacts come together. The usual solution to this problem is to use a multiple-contact structure.

A switch contact can be made entirely from one material, or it can use electroplating to deposit a more suitable contact material. By using electroplating, the bulk of the contact can be made from any material that is mechanically suitable, and the plated coating will provide the material whose resistivity and chemical action is more suitable. In addition, plating makes it possible to use materials such as gold and platinum which would make the switch impossibly expensive if used as the bulk material for the contacts. It is normal, then, to find that contacts for switches are constructed from steel or from nickel alloys, with a coating of material that will supply the necessary electrical and chemical properties for the contact area.

The usual choice of materials is the same as is used for relays (see Chapter 7), with the addition of copper and beryllium–copper.

Switch ratings are always quoted separately for AC and for DC, with the AC rating often allowing higher current and voltage limits, particularly for inductive circuits. When DC through an inductor is decreased, a reverse voltage is induced across the inductor, and the size of this voltage is equal to inductance multiplied by rate of change of current. The effect of breaking the inductive circuit is a pulse of voltage, and the peak of the pulse can be very large, so that arcing is almost certain when an inductive circuit is broken unless some form of suppression is used.

Arcing is one of the most serious of the effects that reduce the life of a switch. During the time of an arc very high temperatures can be reached both in the air and on the metal of the contacts, causing the metal of the contacts to vaporize, and be carried from one contact to the other. This effect is very much more serious when the contacts carry DC, because the metal vapour will also be ionized, and the charged particles will always be carried in one direction. Arcing is almost imperceptible if the circuits that are being switched run at low voltage and contain no inductors, because a comparatively high voltage is needed to start an arc. For this reason, then, arcing is not a significant problem for switches that control low voltage, such as the 5 V or 9 V DC that is used as a supply for solid-state circuitry, with no appreciable inductance in the circuit. Even low-voltage circuits, however, will present arcing problems if they contain inductive components, and these include relays and electric motors as well as chokes. Circuits in which voltages above about 50 V are switched, and particularly if inductive components are present, are the most susceptible to arcing problems, and some consideration should be given to selecting suitably rated switches, and to arc suppression, if appropriate.

The normal temperature range for switches is typically $-20°$C to $+80°$C, with some rated at $-50°$C to $+100°$C. This range is greater than is allowed for most other electronic components, and reflects the fact that switches usually have to withstand considerably harsher environmental conditions than other components. The effect of very low temperatures is due to the effect on the materials of the switch. If the mechanical action of a switch requires any form of lubricant, then that lubricant is likely to freeze at very low temperatures. Since lubrication is not usually an essential part of switch maintenance, the effect of low temperature is more likely to be to alter the physical form of materials such as low-friction plastics and even contact metals.

Flameproof switches must be specified wherever flammable gas can exist in the environment, such as in mines, in chemical stores, and in processing plants that make use of flammable solvents. Such switches are sealed in such a way that sparking at the contacts can have no effect on the atmosphere outside the switch. This makes the preferred type of mechanism the push-on, push-off type, since the pushbutton can have a small movement and can be completely encased along with the rest of the switch.

Switch connections can be made by soldering, welding, crimping or by various connectors or other screw-in or plug-in fittings. The use of soldering is now comparatively rare, because unless the switch is mounted on a PCB which can be dip-soldered, this will require manual assembly at this point. Welded connections are used where robot welders are employed for other connection work, or where military assembly standards insist on the greater reliability of welding. By far the most common connection method for panel switches, as distinct from PCB mounted switches, is crimping, because this is very much better adapted for production use. Where printed circuit boards are prepared with leads for fitting into various housings, the leads will often be fitted with bullet or blade crimped-on connectors so that switch connections can be made.

Fuses and circuit-breakers

All electronic circuits, unless of the microwatt variety powered by a high-impedance battery, must be fused so as to prevent damage that would be caused by excess current. The choice of fuses from the usual bewildering variety is much more strictly governed than the choice of other hardware items, however, because there are often national regulations which must be obeyed if fuses are being replaced or if electronic equipment is being exported. European standards specify fuses of 20 mm length and 5 mm diameter (20×5 fuses), whose specification is outlined by IEC 127. For the USA and Canada, however, the fuse standards are UL198G and CSA22-2 No. 59 respectively, using $1\frac{1}{4}$ in $\times \frac{1}{4}$ in fuses whose characteristics in terms of blowing time and current are quite different from the European standards. These fuses are not of interchangeable dimensions but nevertheless great care should be taken not to mix the types. This is particularly important in the UK where $1\frac{1}{4}$ in fuses which are to the IEC 127 standards on current and time characteristics are in use along with the 20 mm type.

The aim of a fuse is to interrupt current in the event of a fault that causes excessive current to flow. The subject is not nearly as simple as this would suggest, however. If a fuse is used in a circuit in which a high voltage would exist after the fuse blows, then that voltage placed across the fuse might be enough to cause an arc-over, so that current was not interrupted when the fuse blew. Fuses carry a voltage rating, and should not be used beyond that rating, so that a fuse rated at 125 V should not be used in a circuit in which 240 V could exist across a blown fuse. A fuse can be safely used at voltages lower than the rated maximum, but not at any voltage higher than the stated maximum.

The interrupt test rating of a fuse is another value which is not well known. A fuse rated at 1 A might never have to bear a current of 1000 A or more, but its behaviour at such high currents is important. If the fuse continued to pass current, for example, because of conduction by the metal vapour from the wire, it would once again be unsatisfactory. Approval testing for fuses is therefore carried out at very high currents as well as with high voltages across the fuse, and this figure is often quoted.

The quantity that is always quoted for a fuse is the nominal current rating, but precisely what that means depends on the standard to which the fuse has been constructed. A fuse which is rated at 1 A, for example, will not necessarily blow at a current of 1 A, because the blowing of a fuse is a complicated process that involves current and time, and the various standards exist to provide guidelines on the current–time limits. The three main fuse standards that you are likely to come across are the UL198G (USA), CSA22.2 (Canada) and the IEC 127 (Europe, including the UK). These standards are substantially different; a European 1 A fuse would be rated in the USA as 1.35 A, and this is typical over the range of ratings. European fuse ratings are aimed at protection against large overloads caused by short circuits; fuse ratings in the USA aim to provide overload protection from currents that are rather less than short-circuit levels.

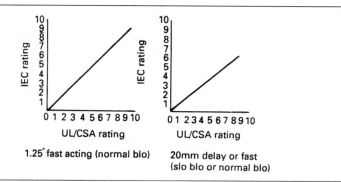

Figure 9.6   Comparison graphs for European and US fuse types.

Very great care must therefore be taken with equipment of overseas origin, or with equipment intended for export, that the fuse types are correct as well as the ratings. Figure 9.6 shows the corresponding ratings of three types of fuses with values on the IEC ratings and on the UL/CSA ratings. Probably the easiest way to remember the conversion is that the fast fuses are rated in the ratio 10/9 (10 A US fuse = 9 A Euro-fuse) and for the others the ratio is 8/6 (8 A US fuse = 6 A Euro-fuse).

Fuses are grouped in five major categories according to their current–time characteristics. At one end of the scale, semiconductor circuits need fuses that will act very quickly on short-circuit conditions. These are now described as super quick-acting fuses, coded FF. These, at ten times rated current, will blow in a millisecond or less, and for twice the rated current the blowing time will be 50 ms or less. The next group comprises the quick-acting class F fuses (classed as *normal blo* in the US), used for general-purpose protection where current surges are unlikely to be encountered. These have a slower blowing characteristic, some 10 ms for ten times rated current and just over 100 ms for twice rated current.

- No fuse, no matter how fast it blows, can adequately protect semiconductors. The use of fuses in semiconductor circuits is aimed as a last resort to avoid damage to other components.

The medium time-lag fuses (type M) will withstand small current overloads that might be caused by charging capacitors. These fuses will blow after about 30 ms on a ten times current overload, and after about 20 seconds on a two times overload. The long time for the smaller overload allows for current surges that are not particularly large but of quite long duration. The time-lag type T fuses (classed as *slo-blo* in the US) will blow in 100 ms for a ten times overload and in about 20 seconds for a twofold overload. Super time-lag class TT fuses allow for 150 ms at a tenfold overload and 100 seconds at a twofold overload.

All fuse ratings are measured at 20°C–25°C ambient temperature, and because the blowing of a fuse is a thermal process, the ratings of a fuse are affected by changes in the ambient temperature. The slower-blowing fuses in particular, which depend on the use of some heat-sinking to delay blowing, need to be derated if they are to be used at high ambient temperatures, and for these types, derating to 60% of nominal value is recommended if the fuse is to be used at 100°C. The derating for the faster-blowing types is considerably less, but in general fuses should not be located at a point in a circuit where high temperatures exist (such as next to a set of power transistors) unless this is done deliberately as a safety measure. Quite irrespective of any derating due to ambient temperature, the normal current through a fuse can be the fuse rated current for a European type of fuse, but not more than 75% of rated current for a US/Canada type of fuse.

Fuses are resistors, and although the larger capacity of fuses have a negligible resistance, this is not true of the smaller types. Fuse resistance is not usually quoted by suppliers, but it can add to the resistance of a power supply and upset stabilization to some extent, although for the larger rated fuses the contact between fuse and fuseholder contributes more resistance in some cases.

Fuse specifications provide a table of currents, in terms of rated current, along with maximum and minimum times for which the fuse should withstand such currents. If you thought that a 1 A fuse would blow when the current exceeded 1 A then you have not been heavily involved in choosing fuses. The US/Canadian specification, for example, provides that a fuse should be able to withstand a current of twice the rated value for more than 5 seconds, and the IEC ratings provide for minimum times for which a fuse should continue to conduct at small overloads, along with both maximum and minimum times for blowing on larger overloads up to ten times rated current. In addition to the normal standard 20 mm or $1\frac{1}{4}$ in fuses, there are fuses both in these sizes and in miniature sizes which have very different current–time characteristics.

Circuit boards

Circuit boards have been of almost a stereotyped pattern until the rise in use of surface-mounted components comparatively recently. The standard board backing materials are either SRBP (synthetic resin bonded paper) or glass and epoxy resin composites, with copper coating. Many suppliers offer such boards with the copper already coated with photoresist, saving considerable time and effort for small batches. These pre-coated boards must be stored carefully, preferably at low temperatures between 2°C and 13°C, and have a shelf life which is typically 1 year at 20°C. The maximum allowable temperature is 29°C.

Board sizes now follow the Eurocard standards of 100 mm × 160 mm, 100 mm × 220 mm, 233.4 mm × 160 mm and 233.4 mm × 220 mm; and there are also the older sizes of 203 mm × 95 mm (8 in × $3\frac{3}{4}$ in) and 304.8 mm × 457.2 mm (12 in × 18 in). Boards can be obtained with edge connecting tongues already in place; these must, of course be masked when the main board is etched. When boards are bought uncoated, photoresist can be sprayed as an aerosol for small-scale production or research and development applications.

- When such boards become out of date and have to be scrapped, you should refer to current practice on disposal to avoid environmental damage.

Many commercial PCBs, particularly for computer or other digital applications, are double-sided, with tracks on the component side as well as on the conventional track side. Where connections are needed between sides, plated-through holes are used. These are holes which have copper on each side and which have been electroplated with copper so that the holes have become partly filled, making a copper contact between the sides. These connections are strengthened when the board is soldered.

The use of double-sided board is particularly important for digital circuits where a single-sided board presents difficulties because of the need to cross leads. The use of a well-designed double-sided board can solve these problems, but care needs to be taken over capacitances between tracks that are on opposite sides of the board. Design and construction of such boards is a skill that is beyond the scope of this book.

A method that has been used with rather variable results in the past is now returning in improved form. The tortuous method of photo-etching a copper sheet can be replaced with the more elegant method of printing the pattern of a circuit directly on to an insulating backing, using silver-based conductive inks. This eliminates the time-consuming photoresist coating, developing and etching steps, and also eliminates the uses of the unpleasant chemicals that are involved.

For experimental uses, circuit tracks can be applied directly by using etch-resistant transfers. Standard patterns include lines of various thicknesses, IC and transistor pads, and pads for mounting connectors, together with a variety of curves, dots, triangles and other patterns. Where the transfer patterns are unsuitable, etch-resistant ink can be applied from a fibre-tipped pen. The standard etching and cleaning processes are then used, and finished boards can be tin-plated for easier soldering, particularly when using flow-soldering machinery.

For a wide range of one-off or development work, however, the production of etched PCBs in this way is too time consuming and stripboard methods are more suitable. The traditional type of stripboards in both 0.15 in and 0.10 in pitch are still available. These are always single-sided, and are suited mainly to small-scale analogue circuitry. For digital circuits there is a range of Eurocard prototyping boards, either single- or double-sided. For connections between strips on opposite sides, copper pins are available which can be soldered to each track, avoiding the difficulties of making soldered-through connections on such boards. Boards can also be obtained in patterns such as the IBM PC expansion card or the Apple expansion card forms.

Surface mounting components and boards are now increasingly featured. Surface mounting is not new; surface mounting boards for amateur use were on sale in 1977, when they were demonstrated under the name of 'blob-boards' at several exhibitions. The developed technique, known as SMT (surface mounting technology), has now spread to professional equipment and has resulted in the manufacture of a whole range of components that are designed specifically for this type of fixing. Components for surface mounting use flat tabs in place of wire leads, and because these tabs can be short the inductance of the leads is greatly reduced. The tabs are soldered directly to pads formed on to the board, so that there are always tracks on the component side of the board. Most SMT boards are two-sided, so that tracks also exist on the other side of the board.

The use of SMT results in manufacturers being able to offer components that are physically smaller, but with connections that can dissipate heat more readily by conduction (although overall dissipation characteristics are generally poorer than their conventionally mounted equivalents), are mechanically stronger and have lower electrical resistance and lower self-inductance. Some components can be made so small that it is impossible to mark a value or a code number on to them. This presents no problems for automated assembly, since the packet or tape of components need only be inserted into the correct hopper in the assembly machine, but considerable care needs to be taken when replacing such components, which should be kept in their packing until they are soldered into place. Machine assembly of SMT components is followed by automatic soldering processes, which nowadays usually involve the use of solder-paint (which also retains components in place until they are soldered) and heating by blowing hot nitrogen gas over the board. Solder-baths are still used, but the hot-gas method causes less mechanical disturbance and can also allow heat-sensitive components to be shielded.

Considerable care is needed for hand-soldering and unsoldering SMT components. A pair of tweezers can be used to grip the component, but it is better to use a holding-arm with a miniature clamp, so that both hands can be free. The problem is that the soldering pads and the component itself can be so small that it is difficult to ensure that a component is in the correct place. Desoldering presents equal difficulties – it is difficult to ensure that the correct components are being desoldered, and almost impossible to identify the component after removal. A defective SMT component should be put into a labelled envelope immediately after removal so that post-mortem testing can be carried out whenever convenient. In some cases, parts may have to be returned to the manufacturer.

Cabinets and cases

The variety of cabinets and cases is as wide as that of the other hardware components. Small battery-operated equipment can be

Table 9.3 The U-value standard cabinet dimensions. Note that for the smaller cases, the E-value is used to denote width

U-number	Height	Width	Depth	(Typical dimensions in mm)
40U	1920	647	807	
34U	1620	600	600	
27U	1290	600	600	
20U	998	600	639	
12U	619	600	639	
6U	230	88*	160	
3U	95	88*	160	

*Smaller cases can be specified as 10E width (38 mm) or 20E width (88 mm)

housed in plastic cases, particularly one-off or developmental circuits, but for the production of equipment for professional use, some form of standard casing will have to be used. As often happens, industry standards have to be obeyed.

The old 19 in rack standard for industrial equipment has now become the IEC 297 standard, with cabinet heights designated as U numbers – the corresponding millimetre measurements are shown in Table 9.3. These cabinets can be supplied with panels, doors, mains interlocks, top and bottom panels, fan plates, and supports for chassis, providing ample space for internal wiring and cooling. Internal chassis in the form of racks and modules can be fitted, usually in 3U and 6U sizes.

Smaller units are accommodated in instrument cases, of which the range is much larger. There is a range of cases which will fit the 19 in units from the standard rack systems so that identical chassis layouts can be used either in racks or in the smaller cases. The more general range of casings cover all sizes from a single card upwards, and also down to pocket calculator sizes. Cases can be obtained with carrying handles for enclosing portable instruments, or for bench or desk use. Many casings can be obtained in tough ABS plastics or in diecast metal form. Metal cases are important where RFI/EMI problems are concerned, and for some types of equipment the casing may need to be designed for minimum emission levels.

Some further degree of standardization is emerging, as far as European equipment is concerned, as a DIN standard 43700 for small cases and boxes along with plug-in modules that fit inside.

Chapter 10

Computer assistance in electronics

Drawing a circuit diagram, analysing the action, and subsequently planning and drawing the silk-screen outlines and the copper track plan for a printed circuit board are activities that are rightly regarded as monotonous and liable to error when they are done by the traditional methods. For a considerable time it has been possible for the larger manufacturer to use computers, mainframe or mini, to aid these actions, but at a cost that would be unacceptable for the small-scale designer or the educational user. The standardization of the PC type of machine and the remarkable amount of computing power that modern examples (using the Pentium or equivalent chips) can achieve now bring computer-aided drawing and PCB planning methods within the reach of virtually any user.

The availability of low-cost computers has made it possible to carry out many tasks in electronics with very much less effort than was once required. At the time of writing, a good PC can be bought for as little as £300, and close scrutiny of the computing magazines will show that some names you might think of as supplying 'basement bargain' computers are, in fact, suppliers of rather overpriced goods. There is not much relationship between price and reliability, and since all PC machines are basically of the same design (and often share the same PCBs) the facilities on the less costly machines are not necessarily less than on the high-priced ones. As in so many other aspects of life, it is often better to avoid famous brand names unless you like paying for a nameplate, and the content of the boxes is more important than their appearance or their nameplates.

In this chapter, the most important contributions of the PC to electronics are described. These comprise Internet access to information and programs, the drawing of circuit diagrams, the analysis of linear circuits (passive or active), the analysis of digital circuits, and the drafting of printed circuit layouts. All are particularly tedious to carry out by traditional methods, and the use of the computer is an immense saving of time, effort and money, particularly to small firms specializing in one-off equipment. The amateur can also gain considerably from these applications because they provide him/her with methods that were once available only to the largest scale of professional users.

The computer

Any computer that is described as being PC compatible can be used for the type of work described here, but the faster modern PC machines are better suited for all tasks, provided suitable software is available. This excludes machines such as the Acorn Archimedes, Atari ST, Apple Macintosh and Commodore Amiga – the use of the letter A is coincidental but a useful way of remembering that these are the incompatible machines, each of which is made by one manufacturer only.

In general, for the type of software you will use you will need a computer whose typical specification is summed up as follows:

Celeron, Pentium or equivalent processor working at 300 MHz or faster
RAM of 32 Mb or more
Hard drive of 6 Gbyte or more
$3\frac{1}{2}$ in floppy drive and fast CD-ROM drive (24× or more)
Fast 56K modem and Internet connection
Colour monitor of 15 in size or more
Inkjet printer

You may feel that the capabilities of the machines currently on offer are much more than you need, but it is by now almost impossible, other than buying second-hand, to obtain computers of lower standard. There is nothing wrong in buying an older machine, but you should be prepared to replace the hard drive when it fails, and this requires some experience in the repair and software formatting of a computer.

One of the considerable advantages of the PC machines is that the scale of production allows important add-ons like hard disk drives to be available at very low prices as compared with other machines. It is almost impossible to have too much hard disk space (or too much RAM), and the depth of your pocket should be the only limitation.

- Note that there is no space in this book to discuss the methods, such as clicking and dragging, used in Windows programs, and the reader is referred to the relevant *Made Simple* book titles for this information.

Internet access

The requirements of a computer for Internet use hinge as much on the modem as on the computer itself. The modem is the hardware item (which uses built-in software) that converts computer signals into tones and vice versa, so that signals can be exchanged using telephone lines that were intended for nothing more than speech. A fairly old computer can be used successfully for browsing the Internet provided that the modem and the browser software will run on that computer. For older machines this rules out the use of the most recent software, such as Microsoft Internet Explorer-5, and since the Internet in its present form has not been around for as long as the PC machine, there is not much software around that can be used on old machines. Older machines cannot make full use of modern fast modems (56K type), so that Internet contact will inevitably be slower. If you have a connection that is virtually free, this is not a worry, but if you are shelling out several pence per minute in telephone charges it can be significant.

- To use the Internet, then, requires software that runs under Windows, and the more modern versions of Windows (95 – version 2 onwards) will provide browser and email software integrated into Windows.

Another point to note is that additional software or information that is needed to make connection is usually supplied by the Information Provider (IP) on CD-ROM, so your computer needs a CD drive to read the software. Some providers will offer the alternative of floppy disks. Many IPs offer free access, so that the only cost is that of telephone time. Others now offer free access (including telephone time) at a fixed rate, typically with a joining fee and a small annual charge.

The hardware that must have priority for any computer used for Internet access is a fast modem. At the time of writing the fastest modems available are the 56K type, but some older computers cannot cope with this modem speed, and even some modern machines can encounter problems, particularly if the operating system is Windows 2000. A machine of the standard described above should be able to cope with a fast modem with few problems.

At the time of writing, the only other options for connections were:

The BT Highway type of connection
A full ISDN link using fibre-optic cable
A link through a TV cable operator

An old computer that is fitted with a hard drive, and running Windows (Version 95 onwards) can probably be used for Internet access. The 80486 type of machine in particular is likely to have enough hard drive space to have a reasonably fast serial port, and to process quickly enough even for Windows software. If all that you want from the Internet is to browse text, download some programs or information, and use email, then you can obtain low-cost software that will be perfectly adequate for these purposes, running under Windows 95, and can run as fast as you need.

If, however, you need to work with Internet sites that make use of elaborate graphics, or of features such as security systems and censorship systems that are included in modem Internet software, then the older type of machine is ruled out. Machines of the older class may already be fitted with a modem, although this might be, by modern standards, a fairly elderly and slow modem. Now that the prices of fast modems have fallen to a reasonable level, you might feel that a machine deserved this enhancement to make it suitable both for fast Internet access and for fax use.

The modern type of computer is likely to be faster, contain a larger size of hard drive, use more memory, and run Windows 98 or its successor, Windows ME. It may also already have a modem fitted so that it is ready for Internet use right away, needing only connection to the telephone line. If you need to buy a modem for this machine then you should use the fastest modem that you can find since this class of machine can do justice to it. The greater facility in terms of hard drive space and in memory size make it possible to use a machine in this class for more than simply Internet access and you can consider using such a machine for all the tasks that you might want to remove from the main machine.

Once you have Internet access, you can obtain information on passive components either by using a search engine (such as Googol, Ask Jeeves or Alta Vista) or by going direct to web sites provided by manufacturers of components. You can also download software such as PSPICE (see later) for working with circuit diagrams and analysis.

Circuit diagrams

The problems of drawing good clear circuit diagrams have haunted electronics engineers for many years, and can now be solved by the use of any of the low-cost CAD (Computer-aided Design) software packages that are now available. One of the most suitable for the purpose is *SmartSketch*, a package that was originally written for Windows 3.1, but which runs equally well on later Windows versions. A later version written for Windows 95 is now available, see the web site:

http://www.intergraph.com/smartsketch/upgrade30.asp

Another excellent option is AutoSketch, a commercial product currently selling for around £90 in its version 7.0. The particular advantage in using AutoSketch as compared with other CAD packages in the same price range is that it is compatible with the

AUTOCAD program that is used by large firms (with a price tag of around £3000 to match). Another version, AutoCAD Light 2000, also costs around £100. Yet another option is the well-known Corel-Draw, currently in version 9. Older versions are still supplied and sell for considerably lower prices; they offer all that you are likely to need for circuit drawing.

- Several of these packages come with symbol libraries that contain electronics symbols, although these are invariably the US symbols.

Using any of the CAD programs you can produce drawings of pro-fessional quality on paper as large as your printer or plotter can handle. Your drawings can include text such as labels, headings and the symbols of mathematics and other specialized applications. If your output is to a laser printer it is sometimes an advantage to specify a resolution of 150 dots per inch rather than the default 300, because this makes lines thicker and allows dotted lines to be seen on the printout.

TYPICAL FACILITIES

Taking various CAD programs to use as examples, we can look at a few of the typical facilities that a CAD package offers. The screen that appears when you run (after configuring) is the *Drawing Screen*, Figure 10.1, which contains the menu bar, scroll bars, and the arrow pointer. The arrow is used for selection and for drawing, with the scroll bars used to perform the equivalent action of sliding a sheet of paper around on the drawing board. All of your drawing work is done on the screen display, but there are several options about the information that is seen on the screen. For example, the grid of fine dots on the screen can be turned on or off using the Assist menu, and the spacing of items is determined from the Set-tings menu.

All CAD programs provide some useful aids to precise drawing. The *Ortho* option allows only horizontal and vertical lines to be drawn, no matter how the mouse is moved. This is useful for electro-nics diagrams in which lines representing connections are either horizontal or vertical. The *Grid* is a more generally useful guide to

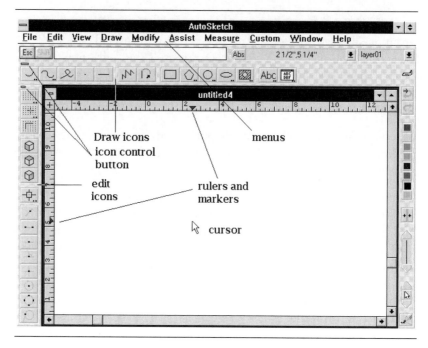

Figure 10.1   The drawing screen of AutoSketch.

positioning the mouse, displaying a grid of faint dots on the screen. The spacing between the dots can be altered to a value that suits the drawing size; 1.0 mm is often useful for electronics work. The Grid is a guide, showing more clearly how far you have moved the cursor, and allowing you to move parts of a drawing into their correct positions, or draw them to correct dimensions.

An *Attach* option allows you to place either end of a line precisely, something that is not possible simply by moving the mouse. It may need to be switched off if you want to draw lines to or from points other than the ends or middle of another line, for example. *Snaps* are used to enforce precision of placing the cursor. When Snaps are selected the cursor can be moved only to snap points, which, if you are working with Grid on, will be the same as the grid size so that your cursor will snap to each grid point.

Your first requirement is to create or buy a set of electronics symbols. If you create your own, you need to work with care to ensure that the size of each component is suitable and that its term-

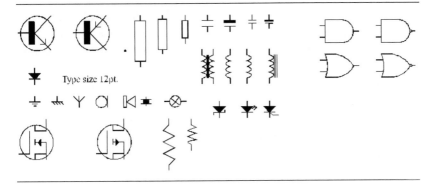

Figure 10.2 A set of circuit symbols, seen in CorelDraw!

inals are on snap points, something that is not always easy for some shapes like transistors and inductors. Unless you have time on your hands, it is often better to buy a set of these sub-drawings. For each shape, whether a small part or a complete drawing, you can rotate, mirror-image and scale the drawings.

An object that is needed in many drawings can be saved as a *Part*, with some point of the object designated as the *Part Base*, a point for manipulating the object. The object can be drawn alone, with nothing else on the screen, and saved in the usual way, or it can be a portion of a larger drawing and saved using the *Part Clip* action. When the Part is needed, the Part option from the Draw menu is selected. This gives a list of all files, with small image (icons) to make selection easier. When the Part is loaded it can be moved to any part of a new drawing.

A simple way of creating a collection of Electronics *Parts* is to draw a circuit with as many types of components as possible, and save each component as a Part Clip. Even better is to collect a number together and save them as a file, Figure 10.2, which can be recalled when you want to create a circuit diagram. The illustration has been taken from the screen, and the printed version of these components is very much better than the screen version because of the limited resolution of the screen.

Any portion of a drawing can be selected so that it can be moved, copied, stretched, or have lines selectively broken. *Move* and *Copy* both make a new copy of an object at another position, but when Move has been selected, the original is deleted. This allows circuits to be built up by copying components from one place to another and then drawing in the connecting lines. The use of Snaps will ensure that all components are perfectly lined up and then the final work consists of copying over the round blobs that mark junctions in the circuit.

Zoomed views allow you to use the whole screen for a view of a drawing, or part of a drawing, at a different scale. This can be used to make the detail of part of a drawing more precise, or to make a small drawing fill the screen, or to make a drawing which is larger than the original limits fit into the screen space. This can be very useful if detail needs to be worked on, and it can be used in the opposite direction if the size of a drawing turns out to be larger than you expected.

Text can be added to any drawing, and the size of text characters is related to the drawing unit – this will almost always need to be changed. There are several sets of characters (or fonts); the default font provides characters that are drawn with simple straight-line strokes.

Because each drawing that is created by CAD methods can be saved as a disk file, it is often possible to make one drawing the basis of another, saving a considerable amount of work, and saving the new drawing under a different filename. Since no drawing need be committed to paper until it is needed, you can adjust as much as you like until the drawing is as you want it, something that is impossible when you work directly on to paper. In addition, where drawings make use of repetitive features (sets of IC pins, parallel lines of a bus, etc.) the multiple-copy actions make the creation of such shapes very fast and easy.

ELECTRONICS SPECIFIC PROGRAMS

The package referred to as *Design Center* or *Eval Center* contains a number of programs that are valuable for drawing and analysing circuits, passive or active. This package can be downloaded from a number of sites, notably

http://sss.mag.com

and various versions ranging from a simple *Student* version to a fairly full implementation can be obtained with no initial payment. The

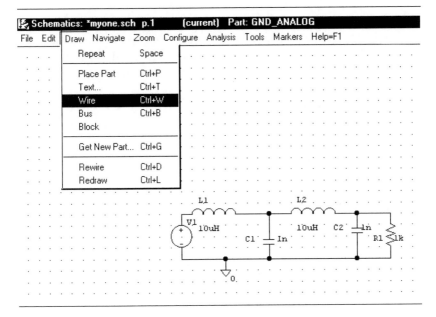

Figure 10.3 A simple drawing of a passive circuit using *Schematics*.

files are compressed (in Zip^{TM} form) and your computer needs to be equipped with the WinZip^{TM} or similar software to expand the files and install them. A particular advantage of using the package is that a drawing made using the Schematics option can then be analysed using the PSPICE option, with graphical output from the PROBE part of the package.

The drawing part of this set is titled *Schematics*, which is specifically intended for drawing electronics diagrams. When you opt to make a new drawing, you can select drawing units for a set of libraries called respectively, ABM, ANALOG, BREAKOUT, EXEL, PORT, SOURCE, SPECIAL, 7400 and FRQCHKX. You can use the *Get New Part* option from the *Draw* menu to select a unit from any of these. The drawing screen displays a set of grid dots, and any component you opt to place will move with the cursor until you click the mouse (a single click will place the item, allowing you to place another elsewhere, a double-click will place the item and end the action.

Figure 10.3 shows a sample drawing made in this way. The input has been selected from the SOURCE set, and the other components from the ANALOG set. The components are dragged into place, and snap to the nearest grid point when you click the (left-hand) mouse button. Each component can be moved by clicking and dragging, and the Draw—Wire command, illustrated, allows you to draw a wired connection together with a dot where one wire joins another.

You are not limited to US conventions (such as the zigzag resistor shape), because you can create symbols for yourself (after some practice) to add to the library. You can also change the default value that each component has as it is entered (such as 10k for a resistor or 1nF for a capacitor) by editing each component in turn. The labelling as V1, C2, R3 and so on, is automatic, and you can determine what form of labelling you want. These features make the use of Schematics much more useful for electronics diagrams than CAD programs. In addition, a circuit created by using Schematics can be analysed by PSPICE and the graph of out/input (or other features) obtained using PROBE.

- Older versions of PSPICE package did not include the Schematics component, and circuit diagrams had to be analysed by node numbering, as noted for ACIRAN following.

Linear circuit analysis

The analysis of linear circuits is based on the principles of the effect of resistive and reactive components on the amplitude and phase of a sine wave. For very simple circuits, this can be done either by drawing phasor diagrams to scale or by the use of algebra to express the circuit impedance as $R + jX$ where X is the reactive component and j is the square root of minus one. These methods are comparatively simple but very tedious, and when the circuit is one that is only slightly more complicated than the most basic filter, the amount of work is enormously increased.

For many standard circuits, the formulae can be obtained from reference books (such as the splendid *ITT Reference Data for Radio Engineers*, now published by Howard Sams & Co. Inc. with the ISBN of 0-672-21218-2), but the amount of manipulation that is required becomes much greater, and in many cases you still have a lot of work to do after working out the results of each formula. The repetitive nature of the work, unless a programmable calculator is used, means that it is very easy to make mistakes.

When the circuit is not a standard one that can be looked up in a reference book, the analysis becomes very much more difficult. It amounts then to combining components, resistive or reactive, in series or in parallel, working out the first two, then combining with the next, and so on until the whole circuit has been covered. The aim is to express the effect of the whole circuit in the $R + jX$ format so that the impedance magnitude (Z) is the square root of ($R^2 + X^2$) and the phase angle (φ) is the angle whose tangent is X/R. This analysis is long, tedious, and very liable to errors. It requires a good grasp of working with complex numbers (numbers including j) and for a large circuit the effort is very considerable. The working for a simple parallel resistor and capacitor will convince you, if you do not already know, that there is quite a lot of work involved.

Another dimension is added when a circuit contains one or more active components. The gain of an active component converts a passive circuit, whose power gain is always less than unity, into an active circuit which can have a power gain of more than unity over a considerable frequency range (the bandwidth). The gain–bandwidth product of the active device needs to be considered, however, as does the effect of the impedances of the active device.

None of this need be a problem for anyone with access to a PC-compatible computer because there are now several programs which ensure that even for very complex linear circuits, the effort of calculating frequency response can be performed by the computer. This brings linear circuit analysis, once possible only on very large and expensive computers, within the reach of any user, amateur or professional.

There are many linear analysis programs available at the time of writing, and if you use the *Schematics* portion of the PSPICE package the logical answer is to use PSPICE itself for analysis, and the companion **PROBE** for producing a graphical output. When you have a circuit diagram, with input and load, you can then produce an analysis, after saving the circuit as a file. The steps are:

1. Run Electrical Rule Check
2. Create Netlist
3. Run PSPICE

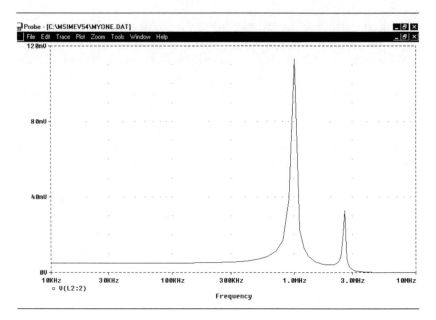

Figure 10.4   The graphical output for the circuit of Figure 10.3 taken at the output.

4. Run PROBE

5. Use Trace to get output graph

You may find that the PSPICE step refuses to run, either because there is an error in the circuit (like no input) or because you have not specified the correct form of analysis (like AC sweep), or because you have not correctly specified the range of frequencies (PSPICE uses k for 1000 and MEG for 1 000 000 – a common fault is to use M, this means 10^{-3}). Figure 10.4 shows a PROBE output graph for the circuit shown in Figure 10.3.

● Note that you can obtain an output at several designated points in a circuit. If your graph shows no changes, it may be because your designated point is the grounded end of a component.

The PSPICE package is ideal if you are using US conventions for circuit diagrams, but another analysis program that is particularly well suited for the small-scale UK user, professional or amateur, is called ACIRAN, and information is available from the website:

http://www.aciran.co.uk/about.htm

ACIRAN has been written by an engineer in the UK and is available in shareware form. For anyone not familiar with this way of distributing software, shareware programs are not obtainable in the usual way from computer shops but by downloading over the Internet and also from specialists in Public Domain and Shareware disks. Although most shareware programs originate in the USA, the ACIRAN program and one other linear analysis program have been written in the UK, so upgrades and support are more readily available.

In addition to its obvious applications to printing tables and graphs of response for a circuit, the use of the ACIRAN program makes it possible to allow for the effect of input and output impedances and of component tolerances, something that is particularly time consuming if done in the traditional ways. This is, however, the type of information that is particularly needed for small-scale production circuit design, so the use of methods based on the computer is a valuable aid to anyone involved in such work. In addition to the conventional R, C and L passive components, transmission lines can be dealt with, a considerable help to anyone involved in UHF design. As well as the built-in systems for dealing with transistors, FETs and op-amps, there is a current-generator component which can be used to simulate the action of any active components much more closely at high frequencies.

ACIRAN cannot be used until it has been notified of the circuit components and connections, and since ACIRAN cannot read a circuit diagram (although if circuit diagrams were prepared with standardized drafting programs such as ORCAD this could be done) it is necessary to use a system for entering the component positions and values. This system depends on identifying and numbering the circuit nodes, as you would when laying the circuit out for construction on PCB or stripboard.

Figure 10.5 shows a typical simple (in terms of components) passive circuit, with input and output and a common earth line, with nodes marked by numbers. A node, in this sense, means a point where components join, and this will normally include the input and the output as well as the earth (ground) connection. Each node can be numbered, and the convention followed by ACIRAN is that the earth (ground) node is always 0 (zero) and the input node is always 1. Other nodes can be numbered as you

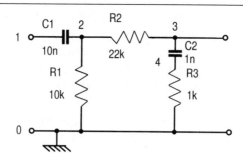

Figure 10.5 A simple circuit with nodes marked.

please, but it is better to take a logical arrangement of node numbers, moving from input to output.

The general rule about nodes is that no node can have more than one number, and nodes are always separated by components. If there is no component between two nodes, one of the nodes must be redundant. If you need to work with nodes connected, use a small resistor value such as 0R33 between the nodes. You can use whatever number you like for the output node, because this will be notified to ACIRAN, but you must use the numbers 0 and 1 for the earth and input nodes respectively. Other linear analysis programs allow you to specify node numbers for all points, but this simply takes longer to set up; since every circuit will have an input and an output there is no reason why these should not be pre-allocated with node numbers.

The total number of nodes in this circuit example is five (numbered 0 to 4). There are also five components, but this is purely coincidental because the number of nodes does not depend in any simple way on the number of components. The point to watch in this illustration is the node numbered 4, because it is easy to overlook this one. Nodes where three or more components join are easy to spot on a well-drawn diagram because of the blob at the junctions, but this type of two-component junction is often less easy to see.

The form of the data can be seen in part in Figure 10.6, a view of the ACIRAN screen when the Data option of the menu has been selected. Small circuits can be checked from this display, but for a complex circuit it is better to print the data out and check against the circuit diagram, which should be numbered to show node

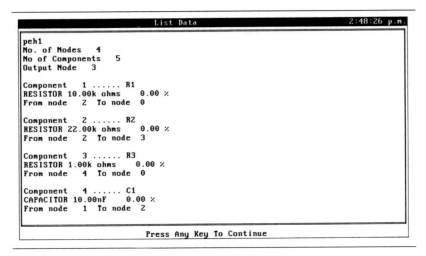

```
                         List Data                          2:48:26 p.m.

peh1
No. of Nodes    4
No of Components   5
Output Node    3

Component    1 ...... R1
RESISTOR 10.00k ohms      0.00 %
From node    2  To node  0

Component    2 ...... R2
RESISTOR 22.00k ohms      0.00 %
From node    2  To node  3

Component    3 ...... R3
RESISTOR 1.00k ohms      0.00 %
From node    4  To node  0

Component    4 ...... C1
CAPACITOR 10.00nF      0.00 %
From node    1  To node  2

                    Press Any Key To Continue
```

Figure 10.6 The components list, showing values and node numbers.

positions. Once this set of data items has been entered, component values can be changed, but you cannot change the connections of a component. If you find, for example, that you have mistakenly entered R3 as lying between nodes 3 and 0 and wish to change this to nodes 4 and 0, you have to do this by deleting the component and re-entering R3 with its correct value between the correct nodes. The list will always show that a component has been deleted as a reminder of what you have done.

The form of entry of values follows the standard methods of specifying value, so that a figure by itself is taken to be in fundamental units of ohms, farads or henries. The suffix values of k, M, m, n, p and so on (lower case or capitals are taken as identical except for M and m) are all recognized, and you can enter values in the format 3K3 if you wish.

The analysis of this circuit is started with a wide frequency range and logarithmic response selected. This is always advisable so that the overall picture of the response can be seen, allowing you subsequently to change the frequency limits and sweep type if you want to see more detail. By specifying a large range initially, you ensure that the circuit has no surprises lurking outside the frequency range for which it is intended. This is unlikely in such a simple circuit, but the point about using ACIRAN is that it allows you to analyse circuits that are far from simple.

Transmission Results			2:48:42 p.m.
Frequency(Hz)	Magnitude(db)	Phase(Deg)	Time Delay(Sec)
1.000E+02	−24.062	85.256	
1.072E+02	−23.465	84.917	−1.314E−04
1.148E+02	−22.869	84.555	−1.313E−04
1.230E+02	−22.273	84.167	−1.312E−04
1.318E+02	−21.678	83.752	−1.311E−04
1.413E+02	−21.084	83.308	−1.309E−04
1.514E+02	−20.490	82.832	−1.308E−04
1.622E+02	−19.898	82.323	−1.306E−04
1.738E+02	−19.306	81.779	−1.304E−04
1.862E+02	−18.716	81.196	−1.302E−04
1.995E+02	−18.128	80.573	−1.299E−04
2.138E+02	−17.541	79.907	−1.296E−04
2.291E+02	−16.955	79.196	−1.293E−04
2.455E+02	−16.372	78.436	−1.289E−04
2.630E+02	−15.792	77.624	−1.284E−04
2.818E+02	−15.214	76.758	−1.279E−04
3.020E+02	−14.639	75.834	−1.273E−04
3.236E+02	−14.068	74.848	−1.267E−04
3.467E+02	−13.501	73.799	−1.260E−04
Press Any Key To Continue			

Figure 10.7   The table of values of magnitude, phase and time delay.

The table that appears following analysis takes the form shown in Figure 10.7 – this is only a part of the table as displayed on the screen. For each frequency in the range, the values of magnitude (amplitude), phase angle and time delay are printed, using units of decibels for amplitude, degrees for phase angle and seconds for time delay. The graph for amplitude is obtained by selecting Graph from the main menu, and appears, Figure 10.8, as the first of a set of at least three.

The other graphs are phase, Figure 10.9, and time delay, of which the time delay is usually of least interest for linear circuit work other than for specialized purposes. When options such as return loss and impedance values have been selected, there will be several more graphs of magnitude and phase angle for each of the other quantities covered.

PCB layouts

Programs for PCB layout have in the past been prohibitively expensive for the small-scale user, although some simplified systems for enthusiasts have been available for some time as shareware. Another choice that is particularly suited to UK professional users is from

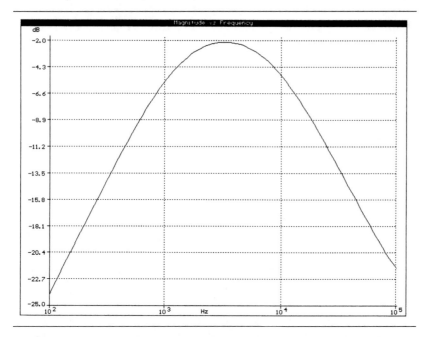

Figure 10.8 The graph of gain plotted against frequency.

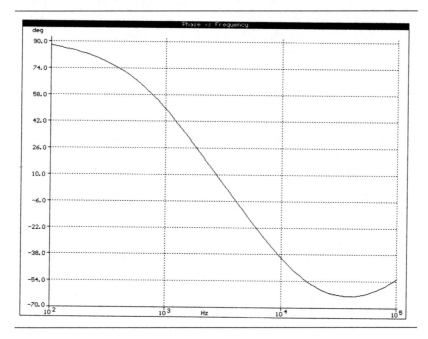

Figure 10.9 The graph of phase plotted against frequency.

Number One Systems whose **EASY-PC** and **EASY-PC** Professional software is the leading software of this type in the **UK**. The address is:

Number One Systems Ltd
Harding Way
St Ives
Huntingdon
Cambs PE17 4WN

Website: http://www.numberone.com

The **EASY-PC** program exists in two forms. The basic form allows you to draw circuit diagrams, linear or digital, using a set of ready-made symbol shapes that you can manipulate on screen and align with lines that are maintained straight and drawn horizontally, vertically or at 45°. You can, of course, add to the symbol library for yourself. The PCB layout can be drawn in the same way, using shapes that are the exact size of electronic components, along with pads and track lines. Although the PCB layout has to be drawn manually, you can manipulate it as much as you need before printing so as to ensure that the printout of the silk-screen view and of the copper track pattern will be perfect. Up to two silk screens and eight copper layers can be catered for.

The Professional version of **EASY-PC** allows the PCB connections to be drawn automatically after the circuit diagram has been drawn, and you can then arrange the components on the circuit board and re-route crossing tracks until the layout is satisfactory, with no need to specify each connection. The program can detect unwanted track crossings and points where there might be a danger of tracks shorting. In addition, when the schematic (circuit diagram) has been drawn using **EASY-PC** Professional, you can analyse the circuit action. Figure 10.10 illustrates a typical example of simple PCB layout.

Linear circuit analysis is carried out by the Analyzer software, providing graphs of amplitude and phase plotted against frequency and digital circuit analysis is carried out by the Pulsar software, providing simulated pulse inputs and showing, as if on a multi-track oscilloscope, output pulses. **EASY-PC** Professional contains both the Analyzer and Pulsar software, limited only in the number of components they can cope with, and you can upgrade if you need to

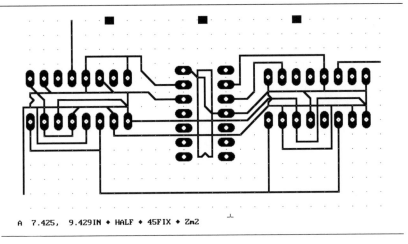

A   7.425,   9.429IN ✦ HALF ✦ 45FIX ✦ Zm2

Figure 10.10   A simple example of layout for a flip-flop circuit.

analyse more elaborate circuits. Keep a close watch on the electronics press (or have your name added to the Number One Systems mailing list) to see upgrade offers.

This, along with the provision (also in EASY-PC) for multilayer boards and file output to Gerber Photo-plotting machines and to numerically controlled drilling machines, makes the professional version more suited for the designer of elaborate circuits who needs to design a large number of boards. Nevertheless, substantial numbers of hobby users of EASY-PC upgrade to the Professional version, and a brief description of this version is included in the book noted here.

For a full description of how to use EASY-PC, see the book (now out of print) *The EASY-PC Handbook* (Sinclair), ISBN 07506 2281 4 (Butterworth-Heinemann).

Index

Troubleshooting Analog Circuits

This book by Bob Pease of National Semiconductor Corporation contains a wealth of practical advice on components, passive and active, and circuits – what makes them work and what makes them fail. With an approach that is miles away from the usual academic text, Bob provides the sort of wisdom that takes a lifetime to acquire and passes it on in a way that you will not forget.

Published by Butterworth-Heinemann

ISBN 0-7506-9499-8